THE BEGINNING OF THE LONG DASH:
A HISTORY OF TIMEKEEPING IN CANADA

MALCOLM M. THOMSON

The Beginning of the Long Dash: a history of timekeeping in Canada

UNIVERSITY OF TORONTO PRESS

Toronto Buffalo London

Canadian Cataloguing in Publication Data

Thomson, Malcolm M., 1908–
   The beginning of the long dash

   Bibliography: p.
   Includes index.
   ISBN 0-8020-5383-1

   1. Time – Systems and standards.   I. Title.

QB223.T56   529'.75'0971   C77-001590-5

The National Research Council
official time signal:
the beginning of the long dash
following ten seconds of silence
indicates exactly one o'clock
Eastern Standard Time

daily radio announcement across Canada,
modified for time zones and daylight time

Contents

Plates

Preface

A century ago, when formal timekeeping was only just getting underway in various towns across Canada, determining the time was a specialized branch of astronomy, and time was measured by the stars. Now, timekeeping is a branch of physics, and time is measured electronically in a laboratory. The Canadian side of that revolution in principles and methods is the subject of this book.

The earliest clocks in Canada, and for long the most widespread, were versions of that most common of astronomical instruments, the sundial. When Champlain built the Habitation at Quebec in 1608 he mounted a sundial on the roof next to the flag of France, unwittingly creating the ancestor of innumerable Canadian town hall clocks. In a predominantly rural society the position of the sun might have remained indefinitely an adequate marker of time. Towards the end of an afternoon, for instance, a farm labourer could see the height of the sun above the horizon as so many handwidths, measured by extending the arm full length and bending the wrist at a right angle, one handwidth representing half an hour. As late as 1840, according to Anne Langton, the household with a watch or clock in Upper Canada maintained a considerable prestige among its neighbours. It was rather in the towns, with the growth of trade and industry, with the rise of telegraphic communications and manufacturing discipline, that mechanically accurate, co-ordinated time was to become important for the general population. The angelus ultimately gave way to the chorus of factory whistles. In Canada, however, our story begins somewhat before this general demand arose and independently of it. In this country the main avenues of activity leading to formal timekeeping were marine navigation, surveying, and scientific research.

Navigation, like surveying, involved determining a position in longitude, expressed as a time difference from a reference point. It had become a much more exact science after the demonstration by Harrison in 1760 of a reliable spring-driven chronometer. With the ship's chronometer set to the local time of, say, Liberpool, the mariner would take observations of the sun or other celestial objects to find the local time. The difference between the observed local time and the chronometer time would then be the longitude difference east or west of Liverpool, and hence of Greenwich. Although scientifically correct, the method was only as reliable as the chronometer which was keeping the time of the home port, and since few ships could afford the cost of an excellent chronometer, navigation was far from perfect. The longer he was away from home port the less reliance could the mariner place on his chronometer. Its time and rate had to be checked at every opportunity, and particularly after a long ocean voyage. There was thus a great need to establish sources of correct time at key locations on the North American continent. Six Canadian harbours were eventually equipped with chronometer timing and rating facilities: Quebec in 1855, Victoria in 1859, Montreal (McGill) in 1862, Saint John in 1870, Vancouver in 1898, and Halifax in 1904 [see Plate 6]. Actually Halifax had had a noonday gun (military) since 1794, and private jewellers provided correct time from 1830. Many communities benefited by the daily sounding of a cannon from a military garrison stationed nearby, and by jewellers who brought the tools of their trade with them.

Timekeeping as an aid to scientific research was undertaken in observatories in Toronto, Fredericton, and Kingston. The Toronto observatory, set up in 1839 by the British government to study fluctuations in earth magnetism, crucial for accurate reading of marine compasses, subsequently became the Canadian headquarters for meteorological research and weather forecasting and a branch of the Department of Marine and Fisheries; astronomical timekeeping was a sideline, though an integral one [Plates 1 and 2]. The Kingston observatory, begun in 1860 as a civic venture, became, like the Fredericton observatory, set up in 1854 [Plate 3], the basis of university courses in astronomy, for which timekeeping was an adjunct.

The early phase of independent local timekeeping was coming to an end by 1883. That year marked the acceptance by many North American railways and major cities of a comprehensive system of time zones, and in Canada a centralization of timekeeping responsibilities and of time signals themselves. The two government observatories at Quebec and Saint John [Plate 5] were placed under the direct supervision of Toronto, mainly to ensure that the time maintained by each of them would be co-ordinated to the nearest tenth of a second. The Saint John daily time signal, distributed by telegraph, became the time standard for the Maritimes. McGill University observatory was by then submitting to regular time comparisons with Toronto, the two signals being combined in 1890 to form the Standard Observer. Because of its location at the heart of the railway system, McGill's daily time signal followed the telegraph across western Canada right out to the coast, where it was used by Victoria and Vancouver.

The Dominion government's great project of surveying the Canadian west had called into being a new survey base at Ottawa, not without objections from the director of the Toronto observatory, and the new centre was inaugurated as the Dominion Observatory in 1905. At Toronto, timekeeping was subsidiary to other activities, and when radio time signals became available transit observations for independent time were discontinued. In contrast, the Dominion Observatory embarked immediately on a program of research into modern techniques of timekeeping and co-operated with the international community of astronomers in an observational program to improve the fundamental star catalogue. In 1936 the government placed responsibility for timekeeping in Canada on the Dominion Observatory.

Our story then becomes a review of the advances in timekeeping adopted, and often pioneered, by scientists at the Time Service in Ottawa. In astronomical equipment, the meridian circle yielded to the smaller broken-type transit, which in turn gave way to the fully automated photographic zenith tube. In clocks, the synchronome pendulum, the most accurate of its kind, was abandoned in favour of the quartz crystal frequency standard, and that in turn eventually could not match the accuracy of the atomic resonator, a technology in which the Time Service, since 1970 part of the National Research Council, is a world pioneer. One result of these developments has been that laboratory time has become more regular than astronomically determined time, so that instead of astronomical checks correcting clocks, the clocks are permitting measurement of irregularities in earth's rotation. The further area of technological transformation we consider, in many ways the most important to the public, is the rise of radio time signals. Radio reception has opened up new opportunities for international comparison of time and longitude; radio transmission has led to the development of the Canadian short wave time signal, CHU.

It has been in adapting to the electronic and atomic age that the skill and ingenuity of scientists at the Time Service have been most clearly apparent. They have pioneered methods in developing a precise cesium atomic standard; developed ways to make the time available by radio and direct telephone line twenty-four hours a day, not only with a minute-by-minute voice announcement in both French and English, but with a code that can be used to correct a clock with millisecond precision. Today, the planes that fly from our airports, the trains that thunder their way along thousands of miles of track, the clock on the Peace Tower that faces in all four directions, and the noon-day guns that roar from St John's to Vancouver, all pay elegant tribute to Canada's Time Service.

Acknowledgments

A work of this nature, though modest in its completed form, actually involves many people. Their help and advice have been generous. Derek Morris, a colleague in the Time and Frequency section of the Physics Division, NRC, kindly read the manuscript and suggested minor rearrangements of material. Prof. J.E. Kennedy has a special interest in the work of William Brydone Jack of Fredericton and helped with that portion of the manuscript. A similar contribution was made with respect to the work of C.H. McLeod of McGill by J. Stewart Marshall; A. Vibert Douglas of Kingston and W.J. Patterson, superintendent of Historic Sites of the St Lawrence Parks Commission, read the pages on Kingston; Willard E. Ireland of the Provincial Archives in Victoria helped with the story of Vancouver and Victoria; and R.W. Tanner, formerly of the Dominion Observatory, read the description of R.M. Stewart's time signal machine.

For the typing and retyping of the manuscript I am indebted to J.S. ('Bunny') Crawford, Shirley Wall, and H.J. Letaif, all of NRC, Ottawa.

Much of the early history has been gleaned from microfilm of the Admiralty records and from the Sessional Papers of the Canadian government, stored in the Public Archives of Canada in Ottawa. It is a pleasure to acknowledge the co-operation so readily extended by several people of this organization. The same co-operation was experienced with officials in the archives at Halifax, Saint John, Fredericton, Quebec City, McGill, Queen's and Victoria, as well as the Harvard University Archives and the National Maritime Museum of London, England. But the manuscript could not have matured in its present form had it not been for the diary maintained by J.P. Henderson during his thirty-seven years at the Dominion Observatory (1919–56), which he freely placed at my disposal.

In addition to those named above, I would like to record my indebtedness to: J.C.L. Andreassen, Archivist, McGill University; Marilyn Arthurs, New Brunswick Museum, Saint John; Rolland-J. Auger, Archivist, Provincial Department of Cultural Affairs, Quebec; Nancy Bignell, McGill Observatory; Jean M. Boone, Harriet Irving Library, UNB, Fredericton; Lyle G. Boyd, Harvard College Observatory; Judge W.M. Cory, Ottawa; Ed Davey, Chatham; Winnifred Dickinson, Vancouver; Clark A. Elliott, Harvard University Archives; Robert Fellows, Provincial Archives, Fredericton; Bruce C. Fergusson, Provincial Archivist, Halifax; R.H. Field, Barbados; L.K. Ingersoll, New Brunswick Museum, Saint John; Jesse Ketchum, Toronto; J.L. Knox, Atmospheric Environment Service, Vancouver; R.J. Lockhart, University of Manitoba, Winnipeg; D.F. MacDermaid, Queen's University Archives, Kingston; Donald C. Mackay, Halifax; J.S. Marshall, McGill University; Keith A. McKinnon, Ottawa; C. Kirkland McLeod, Victoria; A.F. McQuarrie, Atmospheric Environment Service, Victoria; Philippe Mailloux, Montreal; W.L. Morton, Trent University, Peterborough; Fred Phillips, Provincial Archives, Fredericton; George L. Pincock, Atmospheric Environment Service, Toronto; E.F. Porter, Ministry of Transport, Ottawa; J.H. Rector, Rothsay, NB; M. Robertson, New Brunswick Museum, Saint John; Superintendent D.T. Saul, RCMP, Ottawa; Mary Skinner, Atmospheric Environment Service, Toronto; Hon. H.H. Stevens, Vancouver; Morley K. Thomas, Atmospheric Environment Service, Toronto; Don W. Thomson, Department of Energy, Mines and Resources, Ottawa; D.W. Waters, National Maritime Museum, London; Ian Wilson, Archivist, Queen's University, Kingston; and Sidney A. Woods, Victoria.

I should like to thank these individuals and institutions for permission to reproduce the following illustrations: Atmospheric Environment, Department of the Environment 1, 2; Mrs Winnifred Dickinson 15; Earth Physics Branch (formerly Dominion Observatory), Department of Energy, Mines, and Resources 19, 20, 23, 24, 27, 32, 33, 34, 35, 36, 37, 38, 39, 40, 41, 42, 43, 45, 47, 50, 51, 55, 58, 59; J.P. Henderson 29, 48; Malak Photographs Ltd 25, 28, 31, 46, 49; Donald C. MacKay 5; McGill University Observatory 7, 8, 9, 10; NAG Press Ltd, London 4; National Capital Commission 16; National Film Board 26; National Research Council 44, 52, 53, 54, 56, 57; New Brunswick Museum 5; Public Archives of Canada 6, 18, 21, 22, 30; Queen's University Archives 11; Regional Climate Data Centre, Victoria, Department of the Environment 14; Royal Canadian Institute 17; University of New Brunswick Archives 3; Vancouver Public Library 13; Victoria Press Ltd 12.

This book has been published with the help of grants from the Social Science Research Council of Canada, using funds provided by the Canada Council, from the Publications Fund of University of Toronto Press, and from the University of Toronto.

THE BEGINNING OF THE LONG DASH:
A HISTORY OF TIMEKEEPING IN CANADA

1

The origin of
astronomical timekeeping

'A chain of observatories connects the western shores of Ireland and of Spain with the eastern coast of New South Wales, and thus comprises one half of the globe, while the other half is in this respect destitute of any such indispensable means of furthering the purposes of astronomy.' Mr C. Wood, secretary to the Admiralty, was writing on 7 July 1835 to W. Hay, permanent under secretary of state for the colonies. Mr Wood continued: 'Nor do the more immediate practical objects of geography feel less the want of an observatory in that wide interval, as no standard meridian, nor any point to which comparative observations may be referred, exists throughout the whole continent of America. The establishment of a small observatory in Upper Canada would supply this desideratum, which has been so strangely overlooked by the United States, and would thus reflect more honor in this Country.'[1] The observatory at St Helena, he added, was about to be dismantled, and its instruments would be available for such an establishment in Upper Canada.

Lord Glenelg, secretary for war and the colonies, wrote to Sir John Colborne, lieutenant governor of Upper Canada, expressing sympathy with the idea of establishing a small observatory in Upper Canada with the St Helena instruments. When Colborne failed to bring the matter before the Upper Canada legislature, Glenelg wrote the following February to Sir Francis Bond Head, lieutenant governor of Upper Canada (1836–8), requesting him to find out if the legislature of Upper Canada would assume the cost of erecting the observatory or defray its future maintenance. No answer was received, and the instruments remained packed

---

1 Excerpts from the Public Archives recorded by Otto Klotz in *J. Roy. Astron. Soc. Can.*, 12, no. 5 (1918)

away at Greenwich Observatory. Meanwhile, in late 1835 Professor Albert Hopkins of Williams College, Williamstown, Massachusetts, brought back from Europe a sidereal clock, a transit, and other instruments. His example was followed by others, so that soon there were several small observatories established in the eastern United States. Yet as a Canadian astronomer recently noted, 'had it not been for the negligence of Sir John Colborne and the apathy of Sir Francis Bond Head in failing to bring before the legislature of Upper Canada an offer from the Lords Commissioners of the Admiralty, Toronto might have had the first astronomical observatory in the New World.'[2]

Prior to 1835 there was no fixed astronomical observatory on the North American continent, and hence no place where a standard meridian could be drawn. This is not to say that positions were not determined. Men of the Royal Navy who carried on extensive surveys on both the inland and coastal waters of Canada attempted to measure the coordinates of key positions. When Captain (later Admiral) H.W. Bayfield, RN, commenced his survey of the St Lawrence River and the Gulf in 1827, he had already spent nine years on the inland waters of the Great Lakes.[3] Quebec City became his winter headquarters for the next fourteen years. From his temporary observatory on the Plains of Abraham he exercised the utmost care in extending the surveys along to Anticosti, the Magdalen Islands, Prince Edward Island, Cape Breton, and the coast of Nova Scotia as far as Halifax. This was all done by triangulation, supplemented by astronomical observations. Since it was before the days of the cable across the Atlantic, there was no direct link with Greenwich by means of which local time could be compared and difference of longitude determined. There was, however, a link through the heavens in the form of special astronomical tables which gave the position of the moon against the background of stars for every hour of the day in Greenwich time. The position of the moon would be determined by observing the transit time of the moon, together with a group of stars before and after it. Lunar distances, as the method was called, was a way by which one could determine Greenwich time from any location around the world. An occultation, the instant that a star disappears behind the advancing limb of the moon or reappears from the opposite limb, is the most precise method of reading the lunar clock, and does not require a telescope for brighter stars. Occultations were and still are predicted. Transit observations of

2 A. Vibert Douglas in *Queen's Quarterly*, 78, no. 4 (1971)
3 Don W. Thomson, *Men and Meridians*, Vol. 1 (Ottawa, 1966) 186

the stars, on the other hand, yield local time, and the difference between local and Greenwich time is a measure of the difference in longitude from Greenwich. Other phenomena also used to determine Greenwich time, and hence local longitude, were the disappearance of Jupiter's satellites, a transit of Venus or Mercury across the disc of the sun, and a solar eclipse. But these methods were cumbersome even for a land-based observer, and were not at all suited for marine navigation.

In 1839, Harvard College Observatory was founded, with W.C. Bond as director. Bond was not only a skilled technician, a clockmaker by family tradition, but he was also an ardent and painstaking observer. He gathered the observations of the longitude of Harvard dating back to 1743. They involved all the techniques referred to above. In addition, when the telegraph was introduced in the mid-1840s, similar observations made at other United States locations were linked to Harvard. The most advanced technique employed by Bond was the transport of chronometers across the Atlantic. The same observer would check the time and rate of each chronometer at each end of the run, Liverpool and Boston, and derive a value for the difference in local time between these two centres. Great care was taken of the chronometers during the transatlantic voyage. By 1855 a total of 988 exchanges had been accumulated, which resulted in the longitude of Harvard College Observatory being established with greater evidence than any other location in the western hemisphere. It therefore became the fundamental reference point for all coastal and inland surveys of the United States, and ultimately of Canada.[4]

In 1839 Toronto became the site of a magnetic observatory, one of several erected within the British Empire to assist in a global study of magnetic fluctuations, disturbances in the earth's magnetic field. It was established on property belonging to King's College, later to become the University of Toronto. Administered until 1853 by the Imperial Government Board or Ordinance, then until 1871 by the University of Toronto, the Toronto observatory became at that time part of the federal Department of Marine and Fisheries and received growing government assistance to support the development of the new Canadian Meteorological Service.

Lieutenant C.J.B. Riddell, who established the Toronto observatory, brought instruments which included 'a complete set for magnetic work,

4 Otto Klotz reported, in Sessional Papers of the Department of the Interior, 1893, the accumulation of observations used to establish the position of Harvard College Observatory.

transit instruments and the necessary chronometers for obtaining and maintaining correct time, as well as a very full set of meteorological instruments.'[5] When Riddell, because of ill health, retired to England in 1841, he was replaced by Captain J.H. Lefroy, who, except for two years of northern magnetic survey when Lieut. C.W. Younghusband was temporary director, remained with the observatory to the end of the military regime in 1853. An enthusiastic meteorological observer, Lefroy encouraged weather observations to be recorded at guard houses of military establishments in Canada. At the time of his recall to England he was at the point of organizing a system of weather reporting by the Canadian grammar schools.

In 1853 J.B. Cherriman, professor of mathematics and natural philosophy, became temporary director, followed in 1855 by Professor G.T. Kingston, who was appointed by the University to be professor of meteorology and director of the Magnetic Observatory. Through Kingston's energetic efforts the assistance of the schools in making weather observations was organized. In 1871 when federal support of weather forecasting was secured, Professor Kingston was appointed director of the Meteorological Service of Canada in addition to his duties as director of the Magnetic Observatory. Certain locations such as Montreal, Quebec, Saint John, Fredericton, Halifax, Woodstock (Ontario), and Winnipeg became chief reporting stations, manned by competent observers, mainly volunteers who reported their readings each day by telegraph. An additional body of volunteers, including lighthouse keepers, station agents, and others were supplied with instruments and with forms for submitting data on a regular basis. A daily weather report with a forecast of probabilities was started, together with an exchange of data with the United States, and soon the budget was up to $10,000 annually. The service was particularly valuable to shipping.

Time, of course, was an essential element in an observational program which involved co-ordination of observations and production of a synoptic chart showing simultaneous weather conditions across the country each day. At the Toronto observatory, time was maintained from fundamental observations of the stars and the sun; in fact, according to the 1874 report of the director the observatory provided the standard by which clocks and watches in many parts of Ontario had been regulated since 1853. Furthermore the observatory was responsible, as we shall see, for

5 John Patterson, 'A century of Canadian meteorology'

organizing a system of time exchanges by telegraph with other observatories in Canada, thereby assuring co-ordination across the country.

Another influential time service arose in Canada's largest city, Montreal, and like that in Toronto it was ancillary to meteorological reporting. But its growth was different from that of the Toronto time service in several respects. The observatory remained closely associated with a university, McGill, its influence became transcontinental through its adoption by Montreal-based railway telegraph companies, and it owed its origin to the efforts of a private individual. He was Dr Charles Smallwood, an English-born physician with a great interest in meteorology and astronomy [Plate 7].[6] At his home at St Martin, Isle Jésus, nine miles west of the city, Dr Smallwood had assembled, early in the 1840s, a collection of instruments, mainly of his own construction, for observing weather and maintaining time. His small observatory, although provided with a slit in the roof for taking transit observations, was not really equipped for astronomical purposes, because his seven-inch Fraunhofer refractor had to be taken outdoors each time it was used. Smallwood published a number of scientific papers as well as detailed weather summaries. As a member of the Natural History Society of Montreal he met many prominent citizens who had direct contact with McGill University. One was Dr (later Sir) John William Dawson, a Canadian geologist who was appointed principal of the University in 1855. The next year Smallwood was appointed professor of meteorology, but without salary.

It took more than his presence to encourage the construction of the university observatory. In 1858, Mr E.T. Blackwell, president of the Grand Trunk Pacific Railway Company, offered to erect an astronomical observatory if McGill provided a site. Railway companies in the United States were making similar proposals, since they were in need of correct time. A few years later Dr Smallwood offered to bring his meteorological instruments onto the campus, again if a suitable site were provided. In 1862 the work got under way and the stone tower, forming the earliest part of the observatory, was built at a cost of about $2000.

For a decade the elderly Smallwood pursued his several duties with characteristic energy. The time, which was determined astronomically at McGill, became the standard reference for business establishments, the railway, and for the operation of a time ball in the harbour. The Observatory was also connected with the fire alarm circuit of Montreal the bells

6 Nancy Bignell, 'Official Time Signal: 100 Years,' *McGill News*, Montreal, summer, 1962

of which were rung at 7 A.M., noon, and 6 P.M. for the use of workshops and factories. The winding, rating, and correcting of ships' chronometers formed an important item in the work of the McGill Observatory. And commencing in 1869, a signal from Montreal furnished the correct time for firing the noonday gun at Ottawa, for the purpose of regulating the government time under the direction of the postmaster general.

A small amount of revenue was realized by the Bursar's office at McGill from the railway companies, the Harbour Commissioners, and some of the jewellery and other business establishments in Montreal, in return for the time service. Also the observatory received a government subsidy of $500 per annum. But operations always seemed to be severely limited by shortage of funds.

Smallwood died in December 1873. Earlier that year he had signed the diplomas of the first group of McGill engineers to graduate as a class. One of the five young men was Clement H. McLeod, who eventually became professor of geodesy and surveying, and vice-dean of applied science [Plate 8]. As an undergraduate McLeod had assisted in the meteorological work, and so Principal Dawson invited him to take charge of the meteorological observatory and to go to Toronto for a week of special training under Professor G.T. Kingston. Montreal thereupon became a chief station in the new weather reporting network, connected by telegraph to Toronto so that observations could be reported without delay every three hours. McCleod did not immediately assume the timekeeping responsibilities for the city. Instead they were conducted for the next six years by the private observatory of Charles Seymour Blackman, a businessman apparently of some means.[7] When he retired in 1879 and moved to New Haven, Connecticut, Blackman made a handsome gift to the university of all his astronomical equipment, thus launching McLeod on his career as official custodian of time for Montreal and all its connections. In his annual report for 1880 McLeod described the equipment received from Blackwood:

An Equatorial Telescope; a Transit Instrument; a Mean Time Astronomical Clock; a two-day Chronometer; a Sidereal Clock; an Electric-Dial; Two Minute Breakers and an adjustable Meridian Mark with Lens.

The Telescope has a focal length of seven feet, and a 6¼ in. aperture; it is provided with seven celestial, one terrestial and one zenith eye-piece. The object glass is by H.G. Fitz, of New York, and the mounting by W.T. Gregg, of the same

7 Montreal City Directory of 1874

place. The declination circle is 6 inches in diameter, and is graduated to read to minutes of arc. The right ascension circle is of the same diameter and reads to 6 seconds of time. The motions in declination and right ascension[8] are both fast (friction) and slow (tangent screws with handles). The seeker has an aperture of 2 inches.

To accommodate this instrument the tower of the Observatory was partially rebuilt [Plate 9], and a new pier carried up from seven feet below the surface of the ground to a total height of 31 feet 2 inches. The pier is entirely free from contact with the building, and the telescope is thus protected from the vibrations to which a house is subject. The cylindrical form of dome was adopted for the Equatorial Room; it is rotated on iron balls, rolling between two iron troughs or grooves. The lower portion of the tower is used as a clock and electrical apparatus room.

The Mean Time Clock is by Howard, of Boston, with Denison gravity escapement and zinc and steel pendulum compensation. It is provided with electric attachments for minutes and seconds contact.

The Transit Instrument – by Jones & Son, of Charing Cross, London, – is of $3^{1}/_{4}$ inches aperture, and 42 inches focal length [Plate 10]. It is mounted on a solid stone pier, and is sheltered by a small wooden structure. The position of this instrument was of course chosen so as to obtain the best possible view of the Meridian. It is situated about seventy feet to the east of the main Observatory building.

The Sidereal Clock is mounted in the Transit House, where are also an electric dial or seconds jumper and a minute breaker, which are in connection with the mean time clock. A second dial and minute breaker are placed in the equatorial room.[9]

An independent time service arose in another central Canadian town, Kingston, as an offshoot not of meteorology but of astronomy. 'It would appear' later wrote Otto Klotz, the Dominion astronomer, 'that the impulse to erect an observatory in the public park at Kingston, was due to the

8 Co-ordinates in the sky may be expressed in terms of declination and right ascension, which correspond to the co-ordinates of latitude and longitude on the earth. Declination is the angular distance in degrees north or south of the celestial equator. At the equator, which is the projection of the earth's equator in the sky, the declination is zero; at the north pole it is plus ninety degrees, at the south pole minus ninety degrees. Right ascension is the angular distance along the celestial equator measured eastward from the vernal equinox. It may be expressed in degrees from 0° to 360°, or in time from zero to twenty-four hours.

9 Annual Report for 1879 of the McGill University Observatory by C.H. McLeod, Sessional paper no. 9, 1880, of the Department of Marine and Fisheries

interest aroused in astronomy by the annular eclipse of the sun on May 26, 1854, which was observed at Kingston by Lieut. Col. Baron de Rottenberg with a Dollard $2^1/_2$-inch objective, $3^1/_4$-ft focal length, and by Fred J. Rowan with a Troughton & Simms small telscope attached to a transit theodolite.'[10]

Baron de Rottenberg was assistant deputy quartermaster general from 1847 to 1854, and commander of the garrison in 1855. He was one of the scientific gentlemen referred to in a report to the City of Kingston, dated 15 May 1855:

Your committee has much satisfaction in stating that with the valuable assistance of several scientific gentlemen, they will have it in their power to erect during the present summer an observatory near the centre of the ground which will extremely be an ornament to the place; but what is of greater moment, it will have a beneficial effect upon the minds of the rising generation of the city and of the citizens at large, and will add much to our reputation abroad. Many have subscribed specially for the observatory who would not have paid a penny otherwise to the park. Your committee, assisted by the gentlemen aforesaid, inspected the block house on Union Street, found that it is most substantially built, yet without great additional expense, a sufficient degree of solidity could not be obtained. Therefore, your committee abandoned the idea of applying the block house to the use of an observatory.[11]

Klotz noted that 'especially through the effort and contributions of Baron de Rottenberg, Professor Williamson, Judge Burrowes, and Dr Yates, an equatorial of $6^1/_4$-inches was bought for $800 from Mr Clark of Cambridge, Mass. The equatorial was received in the autumn of 1855, and was set up and adjusted on its pedestal under the dome of a small tower erected in the park in the spring of 1856.'[12] The Rev. James Williamson [Plate 11] had been with Queen's University since 1842 as professor of mathematics and natural philosophy. An enthusiastic astronomer he was later to be appointed director of the university observatory. Using the observations of two eclipses, 1845 and 1854, and one transit of Mercury, 1854, Williamson had deduced the longitude of Kingston as 5^h 06^m 08^s.48, the time being taken from a carefully regulated clock with a wooden pendulum. His observations of many transits (or

10 Otto Klotz, *J. Roy. Astron. Soc. Can.*, 13, no. 7 (1919)
11 Report Book v, City of Kingston, 1855. Queen's University Archives.
12 Otto Klotz, *J. Roy, Astron. Soc. Can.*, 13, no. 7 (1919)

'culminations') of the moon in the following decade yielded a value of $5^h\ 05^m\ 54^s.03$, which compared favourably with the 1905 Geodetic Survey value of $5^h\ 05^m\ 52^s.86$.[13]

Building an observatory in the park to replace the small domed tower was beyond the resources of the local group. An application to Parliament for a grant in aid of the objectives of the observatory was successful, and commencing in 1860 Queen's received an annual grant of $500. The observatory was built at a cost of $1,400. A small transit instrument by Simms was purchased for $180, and the loan of a larger and more useful transit, called the Beaufoy, was secured in 1864 on application to the Royal Astronomical Society, London, England. The advantages of the new instrument were described as follows:

The Beaufoy transit, having been properly fixed between its piers, has been carefully adjusted in the meridian and its corrections for level, collimation and azimuth,[14] fully ascertained. The correction for inequality of pivots, which is very small, has also been determined. Having been found, as was to be expected, more stable in its adjustment than the portable transit by Simms formerly placed in the meridian, it is now alone employed for meridional observations, while the latter has been placed in the prime vertical for the purpose of determining the latitudes by a new and very accurate method by means of stars culminating within a few minutes from the zenith. This method is due to my assistant, Mr Dupuis ... and many observations for latitude give a mean of $44°\ 13'\ 21''.7$ N.[15]

As soon as the observatory was placed in operation, Williamson continued to observe lunar occultations, lunar culminations, the satellites of

13 A. Vibert Douglas, 'Canadian Scientists Report – x; Astronomy at Queen's University,' *J. Roy. Astron. Soc. Can.*, 52, no. 2 (1958)
14 Collimation is a measurement made to determine the true direction of the optical axis of a telescope. The geometry of the optical parts can cause the line of sight (the optical axis) to differ from the geometric axis of the telescope tube, a difference called the collimation error. It could cause all the stars to appear systematically too early, or too late.
   Azimuth (azimuth error): A transit instrument is mounted with its axis as nearly as possible in an exact east-west line. If the west end happens to be a bit north of its true position, a south star will appear a little late and a north star a little early. A systematic error of this nature is readily disclosed by comparing the results of north and south stars. A star which passes through the zenith is not affected by azimuth error, and hence is preferred as a time star.
15 James Williamson, 'Report to the Board of Visitors by the Director of the Kingston Observatory for the years 1864–7,' Queen's University Archives 'R.V.R.,' 'Professor James Williamson,' *Queen's Quarterly*, January 1896. Queen's University Archives

Jupiter, and other phenomena within the competence of his equipment. Without the benefit of such refinements as a micrometer or hand key, he was dependent upon the 'eye and ear method' of determining transit times (see below p. 76), and hence of maintaining correct time. This he made available to local jewellers, to the caretaker of the city clock, and to all who desired correct time.

In 1863 Nathan Fellowes Dupuis, an undergraduate, was appointed Observer. A capable technician and a reliable observer, Dupuis subsequently became a professor of mathematics and a dean at the university. Since the cost of an astronomical clock was beyond the budget of the new observatory, Dupuis made one. And when the mean time clock which had been loaned by the principal, the Rev. Dr Leitch, was reclaimed by his estate on his death in 1864, Dupuis replaced it with one of the same design as his sidereal clock. When the city clock was damaged by a fire in 1865, Dupuis made a second face for the observatory's mean time clock, which he mounted on the rear of the case, then backed the clock against the window of the observatory for all to see. From here, gradually, the observatory made correct time available for the city, the port, and the surrounding community.

The importance of astronomy in surveying was shown in the extension of the longitude network through the Maritimes. In the early 1850s when Captain H.W. Bayfield and Commander P.F. Shortland were advancing their surveys of the Gulf of St Lawrence and the Bay of Fundy, the need to adopt a reference meridian for the longitudes of their maps and charts was recognized. Should it be Quebec City or Harvard College? The former had been adopted by the British surveyors, the latter by the Americans, leading to a discrepancy along the Maine-New Brunswick Boundary of more than a mile and a half. In order to preserve uniformity, G.B. Airy, the astronomer royal, agreed with Admiral Sir Francis Beaufort, hydrographer to the navy, that Harvard College be used in preference to Quebec City because there was greater evidence in support of its adopted longitude.

W.C. Bond, director of the Harvard College Observatory, was most receptive to their request for a telegraphic exchange of time and was able to secure financial and instrumental assistance to equip a temporary observatory established by Shortland near the Government House at Halifax. In 1852 the three stations, Halifax, Bangor, Maine, and Cambridge, Mass. (Harvard College), were connected by the Eastern Telegraph line, a total distance of about 770 miles. Bond described the operation: 'The touch of the observer on the break circuit key at Halifax, at the instant of the passage of a star over the wire of his transit instrument, was recorded by the armature magnets at each place; and subsequently the passage of the same star was in like manner recorded for Bangor and Cambridge. A single operation of this kind determined the difference in longitude of these points within half a second of time.'[16] The exercise continued with several such star transits, so that the longitude of Halifax was defined to the hundredth of a second with respect to Cambridge.

In due course the report of Shortland's work came to the attention of Professor William Brydone Jack of the University of New Brunswick in Fredericton. A graduate of the University of St Andrews in Scotland with top honours in mathematics and physics, Jack had accepted a position at Fredericton, as professor of mathematics, natural philosophy, and astronomy, as a temporary posting for a year or two until a more important appointment offered itself elsewhere.[17] He remained at the University of New Brunswick for forty-five years, twenty-five of them as president.

In 1851 he supervised construction of a small observatory of economic design in the hope that it would serve as a useful pattern for any amateur astronomer wishing to build one for himself. Jack was pleased with its efficiency. The college purchased from Merz and Son of Munich a 6-inch equatorial telescope that was for some time considered to be the best in British North America. A transit telescope and chronometer were obtained on loan from the commissioners of the Boundary Survey between Canada and New Brunswick. Many hundreds of careful observations made by Jack marked him as an enthusiastic scientist who enjoyed the practical as well as the theoretical nature of his work. His was the first astronomical observatory to be built in Canada.

When, in 1854, Jack heard how the telegraph had been used to establish the position of Halifax in relation to Harvard College with great accuracy, he wrote to Bond, whose acquaintance he had made some years previously and with whose work he was familiar, requesting his cooperation in determining the longitude of Fredericton with respect to Cambridge, using the electric telegraph method. 'The mean of a vast number of observations which I have made, give, from the calculations in the Nautical Almanac, the longitude of this place 20 seconds of time less than that usually received as correct ... I have depended upon a clock made and regulated to sidereal time by Dr Toldervy, here, and his transit, which is steadier than mine, has been used in all observations of lunar

16 Report of the Harvard College Director for the year 1852. Harvard University Archives.
17 Archdeacon W.O. Raymond, 'The genesis of the University of New Brunswick'

culminations ... The Telegraph Office here is close by Dr Toldervy's observatory so that there would be no difficulty in making a connection with Boston.'[18] Dr James B. Tolvervy, a graduate of London and Glasgow, and now a physician in Fredericton, was pleased to join forces with Jack in this endeavour. The two men collaborated in the time exchange with Bond in January and February of 1855, and found the correction to the adopted longitude of Fredericton to be of the same sign but considerably smaller in size, approximately two seconds instead of the twenty seconds Jack had observed. The measured longitude of Fredericton was 66° 38′ 21″.5 W, ($4^h 26^m 33^s.43$ W). Jack later described some of the details of the operation in an official report to the lieutenant governor of New Brunswick, J.H.T. Manners Sutton:

It was originally intended to have an unbroken telegraphic communication between the Fredericton Observatory and that of Harvard University; but in consequence of the wires from the latter to the office in Boston being out of repair, Professor Bond found it necessary to trust the two excellent sidereal chronometers for the interval, and remarks that 'on examination, I am induced to believe that no greater error has arisen from this source than would have taken place had the communciation been made from the room adjoining the Transit Instrument.'

Professor Bond's chronometers were carefully and repeatedly compared with his Transit Clock, and with each other, both before and after interchanging signals, so as to ascertain their error and rate; and at both Observatories, on each day of operations, the meridian passages of a number of stars were observed in order to obtain the error and rate of the Transit Clocks. But we need not trouble Your Excellency with the tedious details and long calculations connected with this part of the work.

The longitude of Fredericton having been thus verified and ascertained with an exactness much greater than it could have been done by any other method, it might be used as a centre of operations for determining, with equal exactness, the longitudes of all the other places in the Province that are connected with it by Telegraph wires. To do this would be a matter of much importance to the geography of the country in general; and so far as the sea-ports are concerned, it would also be of great service to navigation.[19]

With the encouragement of the governor general, Sir Edmund Head,

18 Jack to Bond, 10 July 1854. Jack Papers, Harriet Irving Library, University of New Brunswick

19 Jack to J.H.T. Manners Sutton, Lieut. Gov. of New Brunswick (c.o. 188, Vol. 124). Public Archives of Canada, Ottawa (microfilm)

Jack and Toldervy in the summer of 1855 commenced their task of determining the longitude of several places in the Maritimes so that the geography of the province might be more accurately known. Their first effort was directed to the important port of Saint John. Here they had the enthusiastic co-operation of William Mills, a well-respected teacher in the Commercial Mathematical School, who had a private observatory on Garden Street near the Roman Catholic Cathedral.

Toldervy in Fredericton and Jack in Saint John corrected their individual clocks from star transits both before and after an exchange of time. The exchange took the following form. Dr Toldervy commenced tapping seconds in synchronism with his chronometer at twenty seconds after the minute, his telegraph key operating a break circuit. The beats continued until the fiftieth second, followed by a ten-second gap, then a single sharp tap on zero of the minute. The sequence was then repeated, the single tap on zero defining the signal, while the succession of taps from twenty to fifty permitted Jack to determine the tenth of a second by which the signal differed from the local chronometer. The sequence was then reversed, with Jack tapping out pulses in synchronism with his clock, from which Toldervy judged time on his local clock to the nearest tenth of a second. It should be noted that when exchanges were made with Cambridge the impulses could be recorded on the spring governor (drum chronograph), an invention by Bond which improved the accuracy of transit observations.

The longitude of Saint John proved to differ but little from the value determined by Admiral Owen. Later Jack learned that Owen had derived his value by transporting chronometers from Boston and back and having them carefully rated by Bond. In effect, then, Jack was verifying earlier work with improved instrumentation. The August 1855 observations gave the position of Mills's observatory as $45°$ $16'$ $42''$ N, $60°$ $03'$ $43''.35$ W (4^h 24^m $14^s.89$ W).

The following year, during July and August, longitude determinations were extended to Grand Falls and Little Falls (later called Edmundston in honor of Sir Edmund Head). Two values had been recorded for these centres, one with Cambridge as reference, the other with the Admiralty value for Quebec City. Jack and Toldervy verified the American measurement. In the ensuing discussion, the astronomer royal, G.B. Airy, said that this result was to be expected, since they had used the American reference meridian. Ideally the observations for the Quebec longitude should have been combined with all the others that were used to define Harvard College Observatory and then the Quebec longitude adjusted. But since the adjustment could not be readily performed at this juncture,

Airy felt that it was preferable for the sake of uniformity to use a single datum, namely, Harvard College Observatory, for the whole continent.

Professor Jack would have liked to become involved in the larger operation of extending the longitude network to other parts of Canada. In a letter to the astronomer royal on 9 October 1857 he stated in part: 'I wish some systematic and uniform plan were adopted for obtaining by Telegraph differences of longitude in the British North American Provinces. In the construction of maps, great and almost irreconcilable discrepancies are now found to exist, as has been experienced by Sir W. Logan in Canada. I would like very much to be engaged in such work.'[20] Doubtless his meticulous attention to detail and his demonstrated scientific ability would have been of great service to that endeavour. As it was, Jack was instrumental in determining the correct position of the time ball observatory at Quebec City, and his measurements were used when the time ball observatory was established in Saint John in 1870. But it was Lieut. E.D. Ashe, RN, of Quebec City who was commissioned, as we shall see, to extend the longitude network in Canada. This he proceeded to do with almost overbearing assurance but somewhat less devotion to scientific detail, his main course of instruction being the few weeks of collaboration with Jack in the November 1855 telegraphic exchange between Fredericton and Quebec. In 1860, W. Brydone Jack became president of his college, newly named the University of New Brunswick, and henceforth was more than ever involved with educational matters. There is no evidence that the splendid equipment of the observatory was ever engaged in the routine observations that would be required for the maintenance of a time service, either on campus or within the city of Fredericton.

Accurate time was of special significance in seaports because of the importance of properly set chronometers to determine longitude in marine navigation. An observatory at Quebec City, for the purpose of providing correct time for the benefit of the shipping of the port, had been advocated since the early 1840s. Captain E. Boxer, the harbour master of Quebec, Sir Richard Jackson, the commander of the forces in Canada, the Council of the Board of Trade of Quebec City, and G.B. Airy, the astronomer royal, had all spoken in its favour. 'In every port which has the same amount of commerce as Quebec' wrote Airy in 1844 'there should be provided means for obtaining time with that security which can be given only by the sanction of official authority. But it is especially

20 Jack to Airy, 9 Oct. 1857. Jack Papers, Harriet Irving Library, University of New Brunswick

desirable in a port where physical circumstances make it so difficult for mariners to conduct successfully the ordinary operations of nautical astronomy for obtaining time.'[21]

When the Admiralty approved the construction of an observatory at Quebec, Airy was requested by the Colonial Office to submit a proposal. The signal was to be transmitted visually, as usual, by a time ball, a large metal sphere prominently mounted on a mast on which it was hoisted aloft to drop precisely at an appointed hour. Airy made several specific recommendations: 'For the visibility of the signal ball, it is necessary that as seen from the ships in the river and from their neighbourhood the ball should be projected against the sky. And for the adjustment of the transit it is desirable but not necessary that some object should be commanded in the north or south direction at a distance not less than two miles. I merely call attention to these points presuming that they have been regarded in the selection of the site.' The pier supporting the transit telescope should be based on solid ground, or on rock if possible. And 'in constructing two temporary observatories I have found very great convenience in making the opening for observing along the length of the ridge of the roof instead of transverse to it. Less difficulty was found in excluding rain (by a simple hinged shutter) than in any other construction which I have seen. I think it probable also that the whole building might be made a little lower (preserving the same elevation of the transit) which appears to be brought in these plans as a desirable object.'[22]

Placed in charge of the planned observatory at Quebec was a British naval officer on half pay, Lieutenant Edward D. Ashe, RN, who had been injured at Valparaiso, Chile, and invalided home in 1849 with a fractured thigh. In the summer of 1850 Ashe was provided in London with several books on astronomy and mathematics selected by Airy and with essential equipment from the stores of the Royal Observatory. These items included 'a 30 inch transit instrument complete, a 42 inch telescope with stand complete, a 6 inch sextant, [and] a shade for artificial horizon.' They remained the property of the Greenwich Observatory, although the two

21 Included in a message from Lord Grey at the Colonial Office to Lord Elgin, governor general, 9 March 1849, for the information of the Legislative Assembly, wherein were copies of communications from the secretary of state and the Royal Engineers' Department, on the subject of erecting an observatory at the Port of Quebec. Public Archives of Canada

22 Report of the Quebec Observatory for the year ended 30 June 1870. Sessional Paper no. 5, 1871. 'having a time ball slide on a mast is not adapted for this climate,' wrote Ashe. He is credited with the ball-in-cage design used at Saint John.

clocks that were needed, once carefully tested in England, were purchased by the Canadian government.

Lieutenant Ashe proceeded to Quebec and, being now in command, characteristically insisted that only the latest design of time ball, namely the one recently constructed for the port of Liverpool, be considered. Airy's only concern was that the machinery should function satisfactorily even in the severe Canadian climate. Major Robinson of the Royal Engineers dispelled all fears in this direction by stating that the vessels all left Quebec when winter set in and returned in spring when winter was over. But even should it be necessary to operate the machinery all winter, the rare occasion that freezing rain or sleet might tend to clog the works could be dealt with on the spot. So the order was placed, and the time ball was shipped to Canada on 3 May 1852.

The Imperial budget contained an appropriation of £526 for the construction of the observatory. It was built during 1854 and 1855 within the citadel on Cape Diamond. The main body of the building was fifteen feet high, fifteen feet deep, and twenty-four feet long; the adjoining tower was forty-four feet high. By 1855 the Quebec observatory was ready to provide correct time for the shipping in the harbour by the dropping of the time ball on the citadel.

The question of the observatory's precise longitude promptly arose. From 1827 to 1841 Captain Bayfield had made Quebec his winter headquarters and the citadel his reference meridian for the coastal survey, the longitude west of Greenwich being determined by moon occultations and eclipses of Jupiter's satellites. A few years later, during the winter of 1844–5, a temporary observatory had been located close to the Citadel at Quebec and astronomical observations made in connection with the boundary problem between Maine and New Brunswick. These earlier estimations were already becoming dated when Ashe heard of the connection by electric telegraph made between Boston and Fredericton and of the greater weight that was accorded Boston over Quebec as a reference meridian on the continent. He wrote to Professor Jack in Fredericton and arranged for an exchange of time by telegraph between that city and Quebec. This was accomplished in November 1855. The first trial between Fredericton and Quebec was not very satisfactory either to Ashe or to Jack, and they recommended that another attempt at longitude determination be made by direct connection between Quebec and Harvard College Observatory, which was later done.

With the longitude of Quebec thus refined, Ashe took part during 1856 and 1857 in the Geological Survey of Canada under Sir William Logan, extending longitudes telegraphically westwards from that city. Ashe commenced his assignment in October 1856 at Montreal, where he set up his portable transit instrument on the roof of the Exchange Building. It was his intention to press his key as the star crossed each wire of his transit, and the resulting click would be heard and timed by his assistant, Thomas Heatley, at the far end of the telegraph line in the observatory at Quebec City. However the roof proved to be unsteady and the observations valueless. Thereafter Ashe did his observing at ground level, making sure to have at least a massive block of stone to serve as a temporary pier. Furthermore, the Montreal experiment required the services of a telegrapher throughout the observing period, which could be upwards of two to three hours. So instead Ashe followed the practice established by Jack and Toldervy of simply comparing clocks by telegraph, then measuring the correction to the field clock by a series of star transits, while the same procedure was followed for the observatory clock back at Quebec. Ashe had a mean time chronometer and a sidereal chronometer, one of which he left at Quebec, the other being used as the field timekeeper. Since the rates of these two involved a coincidence of seconds pulses once in six minutes, they were readily intercompared to the hundredth of a second.

While the initial effort at Montreal must be considered a failure, Ashe's subsequent work was successful. At Toronto in January 1857 he met his friend Professor Kingston, now director of the Magnetic Observatory. Here the facilities of the transit instrument and the sidereal clock of the observatory were made available, and a wire was extended from the telegraph office some two miles away. In February, Ashe went to Kingston, where a temporary observatory was set up in a cross street between Earl Street and Barrie Street. Mischievous boys might have spoiled his efforts had not police come to maintain order. In March, Ashe met Logan in Montreal, and arrangements were made to use the gardener's tool house in Viger Square as a temporary observatory.

Chicago, whose uncertainty of position, amounting to forty miles or more, had been a real concern to Logan, must have presented a unique challenge because of its great distance from Quebec City. Lieut. Col. Graham, topographical engineer, usa, who was stationed at Chicago, assisted by manning the temporary observatory established by Ashe, while Ashe observed at Quebec. For the telegraphic exchange of clock beats, Ashe reports, 'the electric current was transmitted via Toledo, Cleveland, Buffalo, Toronto and Montreal, a distance of 1210 miles, by one entire connection between the two stations, and without any intermediate repetition, and yet all the signals were heard distinctly at either end of the line; the signal occupied only .08 of a second in passing along that distance.'[23] Thankful acknowledgement was made for the help extended by the telegraph companies involved.

In these activities it is interesting that Ashe seemed unaware of a possible error in his work due to his 'personal equation.' The error arises as a result of the observer systematically estimating the instant of bisection a little early or a little late as the star, in its westward drift across the field of view, passes behind each of the vertical lines of the transit's reticule. The relative personal equation of two observers can be determined by having them observe the same program of stars with two instruments at the same location. Alternatively the errors will cancel out if the two observers exchange positions at the two ends of a baseline. As will be seen below the problem was still under active investigation several decades later. Ashe most certainly had a personal equation, but his results may be considered as satisfactory when it is remembered that longitude by telegraph was then in its infancy.[24]

The Quebec observatory's timekeeping duties involved observations of stars and the sun to check the time and rate of the clocks. Each day except Sunday during the navigation season the time ball was released at the citadel for the benefit of ships in the harbour. At 12:45 it was raised to the half-way point, at 12:55 to the top, and at 1 pm local mean time it was dropped. From the adopted longitude of the citadel, 1 pm corresponded to $5^h\ 44^m\ 49^s$ Greenwich time. Some idea of the astronomical side of the Quebec time service, as it eventually developed, can be gained from information for the year 1879. The navigation season, and hence the operation of the time ball, commenced on 13 May and terminated 29 November, with an 8 per cent failure of the time ball. The observations in connection with time giving consisted of 426 complete transits. It was the custom in those days to consider six stars as a complete set. Ashe does not describe a typical operation at the transit instrument, but he doubtless followed the general practice of the day. So 426 transits would constitute seventy-one observing nights.

Over the years Ashe urged government authorities to set up in Quebec a serious astronomical observatory. Officials of the Canadian Institute in Toronto expressed support for his proposal of a National Canadian Astronomical Observatory. In 1864 he was given funds for an eight-inch

23 Geological Survey of Canada, Report of Progress 1857. Geological Survey Library, Department of Energy, Mines, and Resources
24 Otto Klotz, *J. Roy. Astrom. Soc. Can.*, 13, no. 7 (1919)

refracting telescope and a tower; but there was little further public response. Then, in 1872 the Quebec observatory was selected as one of the chief meteorological stations in the network being organized by Professor Kingston in Toronto. Shortly afterwards a new observatory with connecting passageway was built adjacent to Ashe's house two miles from the citadel.[25] The time ball, still on the citadel, was henceforth dropped by electricity, and an example of the sort of problems a time service could involve occurred when Ashe's small daughter pressed the key, permitting the ball to drop one minute early, as she showed her doll how to send a message.[26] By 1877 Ashe's observatory was providing correct time to the Montreal Telegraph Office and to some sections of the Intercolonial and the North Shore Railways.

In the Maritimes the timekeeping centre became Saint John, partly because of its busy commercial port. Before Confederation the town had had several private sources of accurate time. The schoolteacher William Mills, who had come from Ireland about 1836, not only taught mathematics and navigation but also built a small observatory as a hobby. It was this location that was used by Professor Brydone Jack of King's College, Fredericton, when in 1855 he verified the longitude of Saint John by telegraph.[27] Mills also assisted in the operation by contributing some of his own observations, all properly calculated. There were three watchmakers listed in the city directory for 1867. One of these, George Hutchinson, of 70 Prince William Street, claimed that his business dated back to 1819, and his advertisement contained the statement: 'Chronometers rated by a transit instrument and astronomical clocks.'

In 1868–9, the new federal Parliament decided that time ball observatories should be established in Saint John and Halifax in order to assist the mariners in these two busy ports. Not only were they open year-round, but until the Intercolonial Railway was completed in 1876 they were primarily accessible only by sea. The design of the time ball mechanism and observatory at the Citadel in Quebec, already in operation since 1855, served as a model. E.D. Ashe was consulted, and the changes incorporated in the Saint John time ball mechanism to have the ball ride up and down in a cage of four rods may have been on his advice.

Parliament voted $1500 for observatories in Nova Scotia and New Brunswick for the financial year ended 30 June 1869, and a suitable site

25 Sessional Paper no. 5, Department of Marine and Fisheries, 1875
26 Ibid.
27 Report by Jack and Toldervy to the lieutenant governor of New Brunswick, Public Archives of Canada, MG 11 CO 188

was thereupon selected for Saint John. It was on Fort Howe Hill, and was obtained from the War Department at an annual rental of £1 5s. Half the allotted sum, i.e. $750, was spent in Saint John the first year, the other half being reserved for Halifax. The second year a further $390 was spent, finishing the observatory building and the time ball mechanism.

The establishment in Saint John was described in the *Daily Telegraph*, 1 March 1870:

The Observatory on Fort Howe is completed, with the exception of having placed in it the instruments for marking time and taking observations. The building is composed of two wings and is one story in height. The tower is 20 feet high and is surmounted by the iron rods inside of which the ball runs up and down [Plate 5].

The Ball was made by Mr John Kennedy; is of zinc and measures five feet in diameter. It has a piece of $2^1/_2$ inches iron tubing running through the centre across its diameter. The ends of the tube run to the outside surface of the ball and are finished on the outside by a collar which is secured tightly in its place by a screw on the upper side. The four rods which run parallel with each other to a height of 12 feet from the top of the tower, and then are brought together, form a guide for the ball. In the centre of the base of the tower is an iron cylinder 12 feet in length, 16 inches in diameter, which is placed in a vertical position. The cylinder is provided on the inside at the bottom with a rubber cushion, and is opened at the top. A $2^1/_2$ inch piston runs up from the cylinder and passes into the tube at the under side of the ball until it is stopped by a collar cushioned with rubber at a distance of about six inches from the end. The piston is hoisted by a winch and raises the ball 12 feet. As soon as it is raised it is secured in its place by an ingenious contrivance, and at the proper time it is let go by pulling a cord which lets the piston drop, followed by the ball. The ball drops almost instantaneously for a distance of seven feet, where the escape of air from the cylinder is rendered slower and lets the ball down easily to its place ...

Two months later the *Daily Telegraph* reported the appointment of George Hutchinson as director of the Observatory and Time Ball at Fort Howe at a salary of $500 a year. And the paper informed its readers of the operations of the time ball, due to begin the next day: 'At present it is intended that the ball shall be raised half its whole distance of elevation at 15 minutes before one o'clock. At two minutes before one it will be raised to its full altitude, and at one o'clock precisely it will be dropped. The latitude of the observatory is 45° 16′ 43″ North; and longitude 66° West. When the ball drops at one o'clock Saint John time, it will be 5^h 24^m 15^s Greenwich time, the difference between Saint John and Greenwich time being 4^h 24^m 15^s.' Since a caretaker for the building and grounds was paid $104 annually, the expenses of the building and of the two salaries left nothing for the purchase of horological instruments, and until funds could be appropriated for the acquisition of a good transit and sidereal clock the director was obliged to ascertain the time by his own instruments.[28]

Fort Howe, while it commanded an excellent view of the harbour, was too far removed, and the time ball was not easily seen. In 1873 it was therefore moved to the roof of the Customs House on the waterfront, where it was much more centrally located and visible to the general public as well as to the ships in the harbour. In his annual report of January 1876, Hutchinson reported: 'The ball is dropped at the given time, and is taken from a chronometer, which, on each occasion is previously compared with an astronomical clock ... The clock is frequently tested by solar and sidereal observations taken with a transit instrument.' On 20 June 1877 the Observatory and Signal Station were destroyed by the great fire that occurred at Saint John. A temporary ball was mounted on the roof of the Anchor Line Warehouse until the completion of the new Customs Building in 1880, when the permanent time ball was mounted in a commanding position on the north tower. Adjacent to it the small observatory shared the advantage of a firm footing and a rooftop view.

Halifax, unlike Saint John, was more a military base than a commercial port, forming, with Gibraltar and Bermuda, a station in England's Atlantic quadrilateral. But as in Saint John, private individuals furnished astronomical time for the chronometers of commercial ships. In 1832, one William Crawford, a clockmaker who had learned his trade in Glasgow, applied to the House of Assembly for aid to enable him to erect an observatory and to rate and regulate chronometers. He had in his possession a clock and transit instrument belonging to the Marine Insurance Association. He required in addition a reflecting telescope of sufficient power 'to espy the satellites of Jupiter and their various occultations and eclipses; and also a repeating circle.' These would cost £100. With them he would be able to 'regulate ship chronometers so perfectly as to indicate true longitude, with little chance of error, and thus more perfectly guard the property and protect the lives of those who are engaged in foreign commerce.'[29] The investigating committee recommended that the sum be

28 Sessional Paper no. 5, A 1871, Department of Marine and Fisheries, Report of the deputy minister
29 Journal of the House of Assembly of Nova Scotia, 1832. Public Archives of Canada

made available to the Marine Insurance Association rather than to Craw-ford directly.

In 1865 William Crawford was bought out by Robert H. Cogswell, who then enlarged and improved the premises at 155 Barrington Street. Besides dealing in a complete selection of nautical instruments and books and charts, he was credited with being a 'standard authority on true time, keeping Halifax, Boston, and Greenwich time by astronomical clocks, and having a transit instrument conveniently mounted on his premises for the rating of chronometers. He, like his predecessor, has had the rating of the chronometers of the Cunard Mail Steamers of the Bermuda and New-foundland lines since their commencement, and for the general shipping of the port. He has charge also of railway time. He has for years gratui-tously signalled the true time by which the noon-day gun is fired at the Citadel and for several years made up the weather report for the daily evening papers.'[30]

In 1868, as we have seen, the federal government decided to erect and maintain a time ball on Citadel Hill to give true time to masters of sea going vessels, with a sum of $750, half the total appropriated for the two maritime ports, earmarked for Halifax. But in 1869, William Smith, dep-uty minister of the Department of Marine and Fisheries, reported that 'no portion of the money voted for an observatory in Nova Scotia has been expended.' Nevertheless, he continued, 'a site has been offered gratuit-ously on an eligible position on the Dartmouth side of Halifax Harbour by Colonel Hornby. This offer has been accepted and arrangements will probably soon be made for erecting an observatory and time ball on the site alluded to.'[31] A year later, no progress had yet been made. Instead it was decided to defer any action until it had been ascertained whether the time ball observatory at Saint John was of sufficient value to the maritime interests of the port to warrant the expenditure for its maintenance. Furthermore, the private interests of Cogswell may have entrenched themselves so completely within the community that it was thought advis-able not to disturb the status quo. Be that as it may, more than three decades were to elapse before Halifax harbour would be equipped to display official time.

30 G.A. White, *Halifax and its Business*, 90–1
31 Sessional Papers, Department of Marine and Fisheries, no. 11, 1870, 22

2

Integrating the time services

The 1880s saw a national time service beginning to emerge from among the various centres where timekeeping had originated for local reasons. The decade opened with the retirement, due to ill health, of a pioneer, G.T. Kingston, director of the Magnetic Observatory and the Meteorological Observatory in Toronto. He was succeeded by Charles Carpmael, an eminent mathematician who had joined the staff eight years previously. Already rapidly approaching was the transit of Venus, a rare event which occurs when Venus in its orbit passes directly between the sun and the earth. In transit it appears as a small round black object moving slowly across the face of the sun. It aroused considerable popular interest and scientific anticipation. Carpmael, when asked to report to the government on the advisability of Canadian scientists, joining in a worldwide observational program, stated that the practical use of the transit observations was ultimately the more correct determination of longitude. The event was therefore of the greatest interest to all nations, more especially to those largely interested in shipping, adding, 'As another transit will not take place till the year 2004, I should deem it exceedingly unfortunate if the opportunity were allowed to pass without being utilized to the utmost.'[1] Carpmael had the energetic collaboration of the superintendent of the McGill Observatory, C.H. McLeod. Arrangements were made with the astronomer royal for the use of a device which would simulate a transit. A small black round object would move across a bright disc, and the observer would note the time of four distinct events: the first contact of the small object with the leading edge of the bright disc, the moment that the object left the leading edge and commenced to drift across the bright disc,

1 Sessional Paper no. 5, Department of Marine and Fisheries A 1882, Appendix 28

the moment the object arrived at the trailing edge of the bright disc, and finally the instant that the dark object completed its passage across the trailing edge and left the bright disc. Most of the professional men who participated in the campaign did indeed have a dry run with the device, either in Montreal or in Toronto.

The transit was to occur on 6 December 1882. Parliament provided an appropriation of $5000, and arrangements were made for observations at the following places:

- Winnipeg: Professor C.H. McLeod, observer, Mr H.V. Payne, assistant, using a 4-inch achromatic telescope, transit instrument, two chronometers, and other equipment. A temporary instrument shelter was erected for these gentlemen by the Bishop of Rupert's Land on the ground near St John's College.
- Woodstock: Professor Wolverton, using an 8-inch refracting telescope by H. Fitz of New York, its aperture reduced to 6 inches for these observations.
- Toronto Observatory: Professor C. Carpmael, using a 6-inch telescope by T. Cooke and Sons.
- Whitby: Professor Hare, Ladies College, using a 6-inch telescope by H. Fitz of New York.
- Cobourg: Professor Bain, Victoria University, using a 4^1/$_4$-inch telescope by Smith, Beck and Beck of London, England.
- Kingston Observatory: Professor Williamson using a 6^1/$_2$-inch telescope by Alvin Clark and Sons of Cambridgeport, Mass.
- Belleville: Mr Shearman, using a 4-inch achromatic.
- Ottawa: Mr F.L. Blake, observer, B.C. Webber, assistant, using a 4-inch achromatic telescope from McGill University and a transit instrument loaned by the Department of the Interior.
- Montreal Observatory: Professor Johnson, using a 6^1/$_4$-inch achromatic telescope. The object glass of this telescope being found faulty, it was ground over by Clacey of Boston.
- Quebec Observatory: Lieut. Gordon, RN, observer, Mr W.A. Ashe, assistant, using an 8-inch refracting telescope by Alvin Clark and Sons of Cambridgeport, Mass., its aperture reduced to 6 inches for these observations.
- Halifax: Mr A. Allison, using a 4-inch refracting telescope by Dolland.
- Charlottetown: Mr H.J. Cundall, using a 4-inch refracting telescope.
- Fredericton Observatory: Dr W. Brydone Jack, using a 7-inch refracting telescope, reduced to 6 inches for these observations.

The transit observation involved four phases, the arrival and departure of the planet at the east limb, and the arrival and departure of the planet at the west limb of the sun. Since the exact time of each phase as seen at each location was required, a telegraphic exchange of time was made between stations both before and after the transit. C.H. McLeod made a telegraphic exchange with Professor Hough of Dearborn Observatory, Chicago, to verify his longitude. Free use of the telegraph lines during the exercise was a contribution of great assistance.

On the day of the transit the fortunate stations were: Winnipeg, where Professor McLeod got the last two contacts, Cobourg, where Professor Bain got the third contact in an unsteady atmosphere, Belleville, where Mr Shearman got third contact imperfectly, Kingston, where Professor Williamson got second, third, and fourth contacts, and Ottawa, where F.L. Blake also got second, third, and fourth contacts. Considering the time of year and the weather probabilities, the results were quite acceptable, and were considered to be a valuable contribution to the British system of observations.

Canadian astronomy benefited by the surge of popular interest evoked by the transit of Venus. Carpmael was able to secure a six-inch achromatic telescope by Cooke and Sons, which he mounted in the dome of the Magnetic Observatory, with the intention of using it to study solar flares. Meanwhile, a new transit of three-inch objective and thirty-six-inch focal length by Troughton and Simms, a gift of the City of Toronto,[2] improved the timekeeping capabilities, which were soon to be called into service elsewhere in the country.

By 1883 public need for accurate time had increased perceptibly. There had been complaints about the inaccuracy and irregularity of the time ball operation at both Saint John and Quebec. In 1879 Captain E.D. Ashe, director of the Quebec observatory since 1850, found it necessary to refute a statement in the *Montreal Herald* of 20 October that the time ball was not regularly dropped, and not dropped at the proper time. 'I think it quite unnecessary,' he wrote, 'that I should mention that I am quite capable of determining the proper time.' Yet the situation at Quebec had indeed deteriorated. Solar astronomy had not been active for several years, though the annual appropriation to the Quebec observatory was about double the outlay for a comparable service of weather reporting and timekeeping at the Saint John observatory. Ashe had at one time been active in longitude work and also in solar astronomy. But more recently his principal contribution had been limited to the daily reporting by

2 Sessional Paper no. 7, Department of Marine and Fisheries A 1883, Appendix 26

telegraph of meteorological readings and the operation of the time ball during the navigation season. His retirement in May of 1883 provided an opportunity for a reorganization. The time ball observatories at both Quebec City and Saint John were transferred for administrative purposes to the Meteorological Service, under the direct supervision of the Toronto observatory and Charles Carpmael.

Lieut. Andrew R. Gordon, RN, Carpmael's assistant in Toronto, was sent to Quebec to supervise the reorganization. In order to make a complete break with the past, it was arranged for the military authorities to operate the time ball service and report the meteorological readings by telegraph to Toronto three times a day. Gordon transferred the instruments from the Ashe home and adjacent observatory to the Citadel and saw to the necessary repairs. Lieut.-Col. Cotton, the commandant, agreed to 'use all means' to have the duties of the observatory efficiently carried on. One of his officers, Captain C.W. Drury, was placed in charge. The annual appropriation dropped from $2750 to $1100 because of the savings made on the salary and observatory expenses that had been paid to the former director.

When, on 10 January 1886, W.A. Ashe, son of the former director, was appointed to the position, he moved the observing equipment back to the house and observatory on Grande Allée. It may be noted that the city now had four daily time signals, the gun at noon and at 9:30 P.M., the time ball at 1:00 P.M., and the ringing of the fire alarm bells at 1:00 P.M.[3] The telegraph time exchanges which had been instituted under Carpmael's instructions in 1883 between Toronto and Quebec gave assurance that the agreement was within a second.

The new observatory and time ball in Saint John, constructed in 1881, following the destructive fire of 1877, was the centre of astronomical time in the Maritimes. It was not long to remain an independent source. In April 1883 Carpmael wrote to William Smith, deputy minister of marine and fisheries: 'I beg to draw the attention of the Department to the importance of placing the work of giving the time directly under the supervision of this office, for as at present carried on, we have no means of knowing whether accurate time is given at the various observatories. Arrangements are now being made for the absorption of the Quebec observatory by this service, and if the vote for the Saint John observatory was passed through the account for this service and the observer instructed that we had the authority to inspect his work, by exchange of time

3 Report by Andrew R. Gordon to Wm Smith, deputy minister of marine and fisheries, 20 July 1883. Archives of the Department of the Environment, Toronto

signals or otherwise, it would soon be found that there would be no difficulty in securing accurate time at the various places.'[4] He got his wish very quickly. As George Hutchinson, director of the Saint John observatory, remarked in his annual report for 1883: 'On the 1st July last, this observatory was transferred to the Meteorological Department, and was supplied with a set of meteorological instruments, which are read six times in the 24-hours, at regular intervals of four hours, and reports sent to the office in Toronto. [The readings commenced at 3:44 A.M.] The observatory has also been furnished with telegraph instruments, connected by a loop line through the Western Union Telegraph Company's office which are used for exchanging time with Toronto, direct from the observatory, and may also be used in transmitting correct time to any part of the Provinces having telegraphic communication with this city.'[5]

Carpmael, in Toronto, soon noted that Saint John time was frequently in error by more than a second. Inadequate equipment and not the skill or technique of the observer was blamed. A new transit instrument made by Negretti and Zambra of London was acquired during 1884 to replace the old one, which had been used for years, and a sidereal clock was set up to supplement the mean time clock. In December 1885, Hutchinson described his observing technique as follows: 'The method used at this observatory for the determination of the time consists in the observation of from six to ten clock stars, half of the set being observed in a reversed position of the rotation axis of the transit instrument. The error of azimuth is computed from the observation of two pairs of stars, having considerable range and different signs of declination, one pair being observed in a reversed position of the axis. The collimation error is determined by reversing the instrument on slow moving stars, and the error of level by the measurement of the inclination of the axis with the striding level.'[6]

This was the standard observing technique of the day. It was impossible

---

4 Carpmael to Smith, 2 April 1883, Archives of the Department of the Environment, Toronto
5 Sessional Paper no. 7, Department of Marine and Fisheries, A 1884, Appendix B
6 Sessional Paper no. 11, Department of Marine and Fisheries, A 1886, Appendix B.
    Striding level: It is practically impossible to mount a transit telescope with its axis absolutely level, and this means that when pointing to the zenith it may be slightly off to the east or west. The error is measured by means of a level placed from one end of the axis to the other. In some transit instruments, such as the broken-type Cooke, the level could remain slung from one end of the axis to the other without interfering with the operation of the transit, while in others it had to be placed in position for each measurement. In each case it straddled the axis from one end to the other, and hence was called a striding level.

to mount a sidereal clock in the transit room, so the portable mean time chronometer which was used in dropping the time ball was taken into the transit room for each observing session. It was compared before and after with the sidereal clock. There was a systematic error, as the time exchanges with Toronto indicated, and this was attributed by Hutchinson to an error of 'something more than a second' in the adopted longitude of his observatory. He requested that the transit room be enlarged, and also that a drum chronograph be acquired to record the passage of the star at the instant of crossing each wire of the telescope.

The daily routine of the station by 1885 included a signal to the Intercolonial Railway at noon of 75th meridian time, which corresponded to $12^h 35^m 45^s$ local time at Saint John. The railway observed 75th meridian time (or Eastern Standard Time, as it is called today) right through to Halifax. Western Union Telegraph used 60th or Atlantic Standard Time, while the time ball continued to be dropped at one o'clock local mean time, or 1:24:15 P.M. Atlantic Standard Time. Thus three different times were observed within the city.

All four observatories, Toronto, Montreal, Quebec, and Saint John, provided correct time locally for the benefit of industries, watchmakers, and their surrounding communities. Prior to the reorganization, as we have seen, there had been complaints about irregularities in the time service both at Quebec and Saint John, and in order to provide some control Carpmael, with the co-operation of the telegraph company, initiated a system of time exchanges. Clock beats would be requested from each observatory on a notice of only half a day, and these would be recorded at Toronto on a drum chronograph together with the beats of the Toronto clock. The transit observations would also be submitted, and the clock error at each observatory determined. C.H. McLeod of McGill voluntarily co-operated in the exercise. His work was of such a high quality that in 1890 Carpmael combined the Toronto and Montreal time determinations to form the Standard Observer, from which the deviations of each of the four observatories was measured. Initially the time results from Saint John were somewhat erratic, but improved considerably with improved instrumentation. Commencing in 1890, Carpmael took the occasion of the clock exchanges with Saint John to give time checks to the meteorological observer in Halifax. This practice continued until April 1896, when the Saint John daily time signal was available by telegraph throughout the Maritime Provinces.

Carpmael reported in January 1884 that the telegraphic exchange of time could readily be extended at small cost to the country so that every

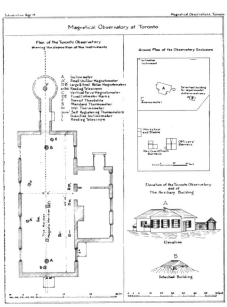

1 / ABOVE The Magnetic Observatory, Toronto, built under the supervision of Lieut. C.J.B. Riddell in 1840
Painting by W. Armstrong in 1952

2 / LEFT Plan of the Magnetic Observatory, Toronto

3 / TOP LEFT   The William Brydone Jack Observatory, built in 1851 on the campus of the University of New Brunswick

4 / LEFT   The time ball on the Citadel, Halifax, Nova Scotia
Drawing by Donald C. Mackay

5 / ABOVE   Fort Howe Signal Station, Saint John, showing the copper time ball which was removed and placed on the Customs House in 1873
Drawing by A.M. Colwell. Courtesy The New Brunswick Museum

NEWFOUNDLAND

Signal Station Latitude and Longitude.	Place.	Signal adopted.	Situation of Time Signal.
47° 34' 10" N. 52° 40' 27" W.	St. John's -	Gun* - - (32 pounder.)	Near Block House on Signal Hill. 517 feet above high water.

CANADA -   -   -   -   -

45° 15' 42" N. 66° 3' 45" W.	St. John, N.B. -	Black ball with gold band.	Northern tower of new Custom House. 123 feet above water. 112 feet above ground. (Drop 12 feet.)
46° 48' 23" N. 71° 12' 35" W.	Quebec - -	Ball - -	At Citadel, 1,370 yards from Observatory. 355 feet above water. 46 feet above ground. (Drop 8 feet.)
45° 31' 0" N. 73° 33' 15" W.	Montreal - -	Black ball - (diameter, 4 ft. 6 in.)	Tower of Harbour Commissioners' building. 145 feet above water. 119 feet above ground. (Drop 8 feet.)
°44 38' 48" N. °63 34 49" W.	Halifax - -	Gun* - -	From the Citadel.

BERMUDA   -   -   -

32° 19' 22" N. 64° 49' 35" W.	Ireland Island	Ball - - (diameter 21 inches).	Flagstaff on western jetty, in front of the main building of the Dockyard. 90 feet above high water. 84 feet above ground. (Drop 30 feet.)

6 / List of time signals prepared from official sources to June 1888 by the Admiralty, London

NEWFOUNDLAND.

Time of Signal being made.		Additional Details.
Greenwich Mean Time.	Local Mean Time.	
h. m. s.	h. m. s.	
3 30 43	0 00 00	Gun fired every day at noon, St. John's (Chain Rock Battery) mean time. [*Note.*—No preliminary signal is given. This signal has been reported to be occasionally in error.]

CANADA.

h. m. s.	h. m. s.	
5 24 15	1 00 00	Ball hoisted half way up as preparatory at 15 minutes before signal. Ball hoisted close up at one minute before signal. Ball dropped at 1^h 00^{m}00^s p.m. St.John mean time.
6 00 00	1 00 00	Ball hoisted half way up as preparatory, at 12^h 45^m 00^s Ball hoisted close up at 12^h 55^m 00^s Ball dropped (by electricity from the Observatory) at 1^h 00^m 00^s mean time for 75° W. long. or 6^h 00^m 00^s from the meridian of Greenwich. If signal fails in accuracy, ball is hoisted half way up, and kept so for half an hour. [*Note.*—Signal not made on Sundays.]
5 00 00	0 00 00	Ball hoisted close up as preparatory at about 5 minutes before signal. Ball dropped (by electricity from the Montreal Observatory) at noon, mean time of the 75th meridian Signal may always be depended upon within half a second. [*Note.*—Signal made during the season of navigation. Signal not made on Sundays.]
4 00 00	0 00 00	Gun fired at noon, mean time of the 60th meridian. [*Note.*—This Gun is fired for local purposes only, and is not to be depended upon for rating Chronometers.

BERMUDA.

h. m. s.	h. m. s.	
4 19 18 3	0 00 00	The ensign usually on the flagstaff is hauled down, and the ball hoisted, as preparatory at 11^h 55^m 00^s a.m. Ball dropped at noon, Bermuda mean time. [*Note.*—Signal made on Saturdays only.]

7 / ABOVE LEFT Dr Charles Smallwood, 1812–73, who founded the McGill Meteorological Observatory in 1862

8 / ABOVE RIGHT C.H. McLeod, 1851–1917, professor of geodesy and surveying, and vice-dean of applied science, who was superintendent of the McGill Observatory for forty years

9 / TOP RIGHT An early view (1881) of the McGill Observatory House with the tower at the southeastern corner. The original tower had been extended upward to accommodate the Blackman telescope.

10 / RIGHT Interior of the McGill transit hut, 1906, showing the Troughton and Simms transit mounted on a solid stone pier

11 / ABOVE James Williamson LL D, 1806–95, professor of mathematics and natural philosophy at Queen's University, 1842–82

12 / TOP RIGHT F. Napier Denison, 1866–1945, meteorologist at Victoria BC 1898–1935, and founding director of Gonzales Observatory in Victoria

13 / RIGHT Nine o'clock gun, Stanley Park, Vancouver BC

14 / TOP LEFT Gonzales Heights Observatory, Victoria BC, built 1914

15 / LEFT T.S.H. Shearman, 1862–1944, meteorologist at Vancouver 1906–15

16 / ABOVE The noonday gun, Ottawa

17 / The birth of Standard Time. Sandford Fleming outlines a plan for universal time reckoning at the Canadian Institute, Toronto, 8 February 1879.
From the original by Rex Woods in the Confederation Life collection of Canadian historical paintings

18 / Edouard G. Deville, 1849–1924, surveyor general of Dominion lands

19 / William F. King, 1854–1916, chief astronomer and first director of the Dominion Observatory

20 / TOP LEFT Otto J. Klotz, 1852–1923, director of the Dominion Observatory 1917–23

21 / LEFT The new observatory, Cliff Street, Ottawa, built in 1912 on the same site as the old observatory. The Alexandra Bridge from Ottawa to Hull, Quebec, is seen in the background.

22 / ABOVE Interior of the old observatory, Cliff Street, Ottawa, built in 1890 on the edge of the precipice overlooking the Ottawa River, showing the transit instrument, observing clock, and collimator

23 / The original straight Cooke transit as used by the Canadian Geodetic Survey. Modified in 1930 to the broken type, it was the principal instrument used by the Dominion Observatory Time Service from 1935 to 1952. The lever on the right raised the instrument on the U frame so that it could be rotated 180° and lowered onto its pivots with the axis reversed.

sea port could be provided with correct time for the benefit of mariners. He also reported 'that the Corporation of the City of London, Ontario, have, in order to avail themselves of the distribution of correct time from this office, erected on the top of their City Hall a time ball signal; they have made arrangements with the Telegraph Company for the use of the wires daily, and the ball has been dropped at noon of standard time, 75th meridian, every day except Sundays and statutory holidays.'[7] A similar service, he noted, could be had by any municipality at the small expense of the wire rental, providing an approved apparatus was installed; but there is no report of other communities making use of the Toronto offer.

The system of time exchanges became increasingly regular. In 1891 the Toronto time service reported the following:

The time exchanges with Montreal, Quebec and Saint John have all been registered on the chronograph at Toronto, the comparisons taking place during the afternoon with Montreal and Quebec, and in the evening with Saint John.

During the year the time at Halifax has also been regularly compared with that at Toronto, the comparisons taking place during the same evening as that with Saint John.

The errors of the Toronto clock, and of the time-pieces used by the observers elsewhere, are computed from the latest observations.

The examination of the monthly clock and chronometer comparisons and transit observations, sent in from the observatories at Quebec and Saint John, has been performed.

During the year observations of 783 stars and 1 solar observation were made from which the time at the Toronto Observatory was obtained. The position of the stars used in the reductions are from 'Berlin Jahrbuch.' The collimation error of the transit instrument has been determined frequently from micrometrical measurements on the collimating telescope and by reversals on 'Polaris' and other stars.

The following table shows the difference between the time by standard observer and that given at the various exchanges.

The sign + indicates that the time as sent from the various observatories is faster than that by the standard observer.

The time by standard observer is obtained by taking the arithmetical mean of the times as determined at Toronto and Montreal, after applying the personal equations between the observers and the director of the Magnetic Observatory, whose absolute equation is known to be almost insensible.

7 Sessional Paper no. 7, Department of Marine and Fisheries A 1884, Appendix 36

	Toronto	Montreal	Quebec	Saint John
1890	Secs.	Secs.	Secs.	Secs.
October 8th	+0.06	−0.06	−2.15	−1.52
do 31st	+0.27	−0.27	−1.23	−1.36
November 21st	+0.10	−0.10	−0.15	−1.53
December 10th	+0.68	−0.68	.....	.....
do 30th	+0.23	−0.23	+3.70	−1.78
1891				
January 21st	+0.18	−0.18	−1.32	−1.63
February 6th	.....	.....	−3.02	−0.74
do 25th	+0.36	−0.36	+0.03	.....
March 17th	+0.11	−0.11	+0.11	+0.31
April 3rd	0.00	0.00	+0.54	+0.08
do 20th	+0.29	−0.29	0.00	−0.20
May 7th	+0.28	−0.28	−2.10	.....
do 21st	+0.15	−0.15	+0.03	−0.06
June 9th	+0.15	−0.15	−0.38	.....
do 25th	+0.15	−0.15	+1.45	−0.36
July 9th	+0.09	−0.09	−0.67	−0.93
do 31st	.....	.....	−0.07	−0.71
August 27th	+0.23	−0.23	+0.77	−1.28
September 11th	+0.11	−0.11	−0.08	−1.67
October 5th	+0.04	−0.04	−0.08	−1.39
do 30th	+0.01	−0.01	−1.39	.....
November 13th	+0.34	−0.34	+0.53	−0.99

Note – Where no exchange has been made with Montreal the Toronto time corrected for its observer's personal equation is adopted as standard time for the comparisons with Quebec and Saint John.[8]

The development of communications between centres spread across a country of such broad east-west extension was making acute the confusion arising from a multiplicity of local times. The pressure for standardized time zones was, by the 1880s, becoming irresistible. The issue was being well aired in the United States. In 1869, Charles F. Dowd, principal of Temple Grove Ladies Seminary in New York State, had spoken out as champion of a system that would bring order out of chaos in the time-

8 Sessional Papers, Department of Marine and Fisheries, 1892, 127

tables of the American railways by reducing the multitude of local reck-
onings from community to community to an orderly array of four time
zones in the United States.[9] A decade later Cleveland Abbe (1838–1916),
the 'father of American meteorology,' presented his important paper on
Standard Time in 1879 before the American Meteorological Society. He
strongly recommended the adoption of a uniform time for the whole of
the country, based on local mean solar time of the 90th meridian.

In Canada the foremost spokesman was Sandford Fleming, chancellor
of Queen's University. One of Canada's leading civil engineers, Fleming
had found the question drawn forcibly to his attention in 1876 when an
error in printing P.M. for A.M. in a railway timetable in Ireland caused him
to miss a train connection, with subsequent embarrassment to himself and
his friends. When, in 1880, at the age of 53, Fleming retired from active
engineering and was elected to the position at Queen's University which
he held for the next thirty-five years, he enjoyed the freedom to devote
himself to the study and support of great causes that would benefit
Canada and the Empire. The forum he used for his discussions were such
scientific, engineering, and cultural societies as the Canadian Institute,
the Royal Society of Canada, the American Meteorological Society, and the
American Society of Civil Engineers. And of his many causes the one that
earned him perhaps the most international recognition was time reform.

The proposal, which had been advocated on a national or regional basis, was seen by Fleming to be one of broad international significance.[10]
Yet in 1878 he was denied a hearing when he wished to present a paper
before the British Association for the Advancement of Science on the
ground that his ideas were utopian. The following year, as a result of a
paper presented to the Canadian Institute entitled 'Time Reckoning and
a Prime Meridian common to all Nations,' a memorial was sent from the
Canadian Institute of Toronto to the governor general for submission to
various scientific societies and foreign governments [Plate 17]. In replying
to this memorial on 18 June 1879, G.B. Airy, the astronomer royal,
observed that 'it has been the custom of Her Majesty's Government to
abstain from interfering to introduce novelties in any question of social
usage, until the spontaneous rise of such novelties has become so extensive as to make it desirable that regulations should be sanctioned by
superior authority.' But, nevertheless, 'it appears desirable that the question should be extensively ventilated by the memorialists, and should be

9 Charles N. Dowd, 'Charles F. Dowd'
10 Sandford Fleming, *Uniform Non-Local Time*

submitted by them to the principal Geographical and Hydrographical bodies, including (perhaps with others) the Royal Geographical Society, and the Dock Trustees, or other commercial bodies, at London, Liverpool and Glasgow.'[11] The idea was no longer thought to be inconceivable.

The international campaign for the adoption of Standard Time, as the new system of time zones was called, was greatly encouraged by the successful demonstration of its use by the railway companies of the United States and Canada who adopted it on 18 November 1883. Many Canadian communities made the change at the same time, and no doubt notices such as the following in one form or another were to be seen in many places: '... standard time, 17 minutes in advance of solar time, which will be adopted in the College commencing Monday next, the 19th inst. University College, Toronto, 16 Nov. 1883, President Daniel Wilson.' Not all Canadians favoured the new idea. Major General D.R. Cameron, commandant of the Royal Military College at Kingston, fought a valiant rearguard action in 1891 as he sought to stave off passage of a bill through Parliament to legalize standard time. The bill received first and second reading in August 1891, but was not passed. Except for matters of national concern, legislation on time has remained in the provincial domain.

A less principled sort of opposition to standard time was recorded in the Ottawa *Free Press* in 1893: 'London, Ontario, May 31 – Today Police Magistrate Parke gave a decision which is of interest to every holder of a liquor license in Ontario. In other words, it was His Worship's judgment in the test cases against C.W. Davis and Jas James, who purposely kept their bars open till 10 o'clock solar time the other night, which is nearly half an hour later than standard time. The judgment was a short one. The magistrate simply said that all the authorities were in favor of solar time, and without some act to legislate standard time the other must govern. He therefore dismissed the case. The effect of the judgment will be to allow bars to remain open nearly half an hour later Saturday nights, as well as every other week night.'

During these years, as a result of the campaign in which Sandford Fleming played a prominent role, the idea of a worldwide system of time zones was gradually gaining acceptance. Two main principles formed the basis of Fleming's discussion of time reform. 'Whatever system might be adopted', he later recalled in his presidential address before the Royal Society of Canada in 1890, 'it was felt that it should be on the fundamental principle that there is only one time. It was moreover held to be that there

11 *Supplementary Papers*, Royal Astronomical Society, 1880.

should be only one reckoning of time common to all nations; and to secure a common reckoning, one established zero and one common unit of measurement became necessary.' An important step was represented by an international conference of twenty-five nations held in Washington in 1884 to consider the adoption of a zero meridian. After a month of debate, the conference resolved, by a nearly unanimous vote, to recommend 'the adoption of the meridian passing through the centre of the transit instrument at the Observatory of Greenwich as the initial meridian for longitude ... That from this meridian longitude shall be counted in two directions up to 180 degrees, east longitude being plus and west longitude minus ... The adoption of a universal day [which] is to be a mean solar day [and] is to begin for all the world at the moment of mean midnight of the initial meridian, coinciding with the beginning of the civil day and date of that meridian; and is to be counted from zero up to twenty-four hours ... That as soon as may be practicable the astronomical and nautical days will be arranged everywhere to begin at mean midnight.'

On 1 January 1885 the twenty-four-hour system was adopted at the Greenwich Observatory, the seat of control for all the public clocks of Great Britain. In a circular issued by the Canadian Pacific Railway in June 1886 the twenty-four-hour system was officially adopted for use on the company's lines. Despite the action of the Washington Conference, however, the nations were slow to take action in the matter of the adoption of a prime meridian common to all. As Fleming had foreseen, national jealousies stood in the way of the general acceptance of Greenwich. Nevertheless, the agitation had been helpful in creating a recognition everywhere of the importance of agreeing upon a universal prime meridian. It was with some degree of satisfaction that Fleming was able to tell the Royal Society of Canada in 1890 that 'without taking into account Central Europe, where the reform is on the eve of adoption, the unification of time-reckoning has so far advanced that in Japan, Norway, Sweden, England, Scotland, Canada and the United States, all well regulated clocks strike the hours at the same moment (although locally the hours are distinguished by different numbers), and the minutes and seconds in all these countries are absolutely synchronous.'[12]

In Canada the degree of success in standardizing time zones was summarized in 1897 by the director of the Meteorological Service:

Nova Scotia – The Railways and Telegraph Companies use the 75th meridian or

12 L.J. Burpee, *Sandford Fleming*

(eastern standard time) but for other purposes the time generally used is the 60th meridian or (Atlantic standard time). Example: Halifax 5 A.M. = 9 A.M. GMT.

New Brunswick – The Railways and Telegraph Companies use the 75th meridian or (eastern standard time) but most places in this Province use local mean time. Saint John local time is 35m 44s fast on 75th meridian or (eastern standard time).

Ontario and Quebec – (eastern standard time) 75th meridian is generally used for all purposes throughout these Provinces as far west as Port Arthur [now Thunder Bay]. Quebec, Montreal, Ottawa, Toronto – 4 A.M. = 9 A.M. GMT.

The Railway and Telegraph Companies east of Port Arthur, as far as the Atlantic coast, use 75th meridian time or (eastern standard time).

Manitoba, North West Territories and British Columbia – West of Port Arthur to the Pacific Coast, the Railways, Telegraph Companies, Cities, and Towns all use the same time, according to the different meridians hereunder mentioned.

From Port Arthur to Brandon, Manitoba, the time used is the 90th meridian or (central standard time). Winnipeg – 3 A.M. = 9 A.M. GMT.

From Brandon to Donald BC the 105th or (mountain time) is used. Calgary 2 A.M. = 9 A.M. GMT.

From Donald westward in British Columbia, the 120th or (Pacific time) is used. Vancouver and Victoria 1 A.M. = 9 A.M. GMT.

Also west of Port Arthur to the Coast, the Railways use the 24-hour system, the P.M. hours from noon to midnight being numbered from 12 to 24.[13]

Because of its importance the recommendation of the Washington Conference concerning the beginning of the day was also the subject of further action. The Astronomical and Physical Society of Toronto, forerunner of the Royal Astronomical Society of Canada, joined forces with the Canadian Institute in 1893 in a recommendation that the astronomical, the nautical, and the civil days be unified. The first two commenced at noon, except that the nautical day was ending as the astronomical day was commencing. Hence there was always the danger of confusion in reading the navigation tables, with resultant possibility of tragedy and loss. This recommendation fell on deaf or hostile ears, particularly at the US Nautical Almanac Office. It was not until 1925 that the astronomical day was adjusted to conform to the Greenwich civil day by changing zero hours of 1 January 1925 astronomical time to 12 hours. Since then the astronomical and nautical day have conformed to the civil day.

13 R.F. Stupart to Major F. Gourdeau, deputy minister of marine and fisheries, 21 Dec. 1897. Archives of the Department of the Environment, Toronto

While the growth of international communications was leading to a co-ordination of timekeeping around the world, it was also bringing cities and countries into clearer relation with each other through refinements in longitude determination. The most significant Canadian developments in this regard occurred in Montreal, because that city was the centre of the country's survey network. The longitude of the McGill Observatory was still, in 1883, based on the work of E.D. Ashe in 1857. C.H. McLeod recognized the possibility of error arising from the fact that no attempt had been made to determine the personal equation in the link between Quebec and Harvard College Observatory. The suspicion of error was further heightened when in 1881 General Cutts of the United States Coast and Geodetic Survey used a site on Mount Royal as a heliotrope station in a geodetic net. By using a mirror, and directing a beam of sunlight in turn towards the stations in Dannemora in New York State and Bellevue in Vermont, the longitude of the Mount Royal station was triangulated to be $4^h 54^m 21^s.68$. The measured distance to the observatory from the heliotrope station of $2^s.81$ gave a value of $4^h 54^m 18^s.87$, which was more than a second at variance with the value established by Ashe.

McLeod felt responsible for the accuracy of the Montreal position, because of its importance as the reference for the Canadian survey. Also, the time service of the observatory was receiving careful attention. Observations were being made on every clear night to obtain a close control of the clock rate. The time ball in the harbour was dropped with but few failures, and McGill time was enjoying wide recognition throughout the city and across the telegraph networks of the railway companies. The importance of accuracy in the time service for the use of the large number of ocean ships visiting the port of Montreal was recognized. Furthermore, this accuracy was based on the precision with which the geographical position of observatory was known. An uncertainty of a second of time was intolerable.

Arrangements were therefore made in 1883 with Professor E.C. Pickering, director of the Harvard College Observatory, for the redetermination of the longitude of McGill Observatory. Free use of the telegraph lines of the Great Northwestern Telegraph Company was arranged for by Angus Grant, manager of the Montreal office, an example of the good rapport that existed between McLeod and men of the business world. The work was done by Prof. W.A. Rogers of Cambridge, assisted by J. Raynor Edmonds, while at Montreal C.H. Chandler provided the necessary support for Prof. C.H. McLeod. Telegraphic exchanges of clock signals and observations for the determination of clock errors, both before and after each exchange, were made on six nights, and a complete time determination for personal equation was made by both observers (Rogers and McLeod) on two nights at Montreal and two nights at Cambridge. Half the expense of the exercise was born by Harvard.[14]

When the calculations were completed, McGill College Observatory was found to be west of the centre of the dome of the Harvard College Observatory $9^m 47^s.510 \pm {}^s.019$. The pier of the transit instrument at the McGill College Observatory was therefore in longitude $4^h 54^m 18^s.543 \pm {}^s.045$ west of Greenwich.[15] This meant that the discrepancy with the observations of 1857 from Quebec had been reduced, but remained $0^s.803$. McLeod then collaborated with Charles Carpmael in a similar venture which yielded the position of the Toronto transit as $5^h 17^m 36^s.649 \pm {}^s.049$.[16] Thus, at the end of 1883, Canadian longitudes were still based on Harvard College.

Although the laying of transatlantic cables in 1866 and 1869 had led to further improvements in the longitude determination of Harvard College Observatory vis-a-vis Greenwich Observatory, doubts developed concerning the accuracy of the methods employed. By 1890 the feeling had arisen that there ought to be an independent determination for Canada by direct transatlantic cable connection with Greenwich. The matter was brought to the attention of the British government at the request of the Royal Society of Canada through the governor general. Sir Charles Tupper, Canadian high commissioner in London, also placed the question before the astronomer royal, who gave it his hearty support. Similar proposals for improvement of longitude determinations had been advanced by the superintendent of the United States Coast and Geodetic Survey, and on behalf of the European longitudes by Otto Struve, director of the Pulkovo Observatory in Russia.[17]

The memorial on longitude addressed by the Royal Society of Canada to the minister of marine in May 1890 ran as follows:

1 Now doubt has recently been thrown on the accuracy of the result of observations by which the longitude of Harvard Observatory has been obtained. This doubt, of course, affects the positions of all places determined by reference to it – that is to say, briefly, it affects the whole geography of the continent. As there are

14 Sessional paper no. 7, Department of Marine and Fisheries A 1884, Appendix 37
15 W.A. Rogers and C.H. McLeod, 'Longitude of McGill Observatory'
16 Charles Carpmael and C.H. McLeod, 'Longitude of the Toronto Observatory'
17 C.H. McLeod, 'Report to the deputy minister of marine, 1890,' Sessional Papers, Department of Marine, 1891, 75

better means available at present for observations and interchange of signals across the Atlantic than at the time of the American determination, it is deemed of great importance that an effort should be made at once to remove the doubt referred to.

2 The Department of Marine, more particularly, is interested in the work, as it affects navigation. The accurate determination of a ship's position at sea, and therefore often the safety of the ship depends on the chronometer. The error of the chronometer has always to be determined in leaving a Canadian port by reference to the local time, and the longitude of the place to Greenwich. This Canadian longitude again is determined by reference to the longitude of the base station, such as Montreal or Harvard Observatory, hence the necessity for extreme accuracy for the base station.

3 The object to be attained is not only of Canadian but of Imperial and not only of Imperial but of International importance.[18]

The Admiralty accordingly set aside the sum of £350 for the instruments and £300 for the operations connected with the work. The sum of $2000 was appropriated by the Parliament of Canada. The details were arranged by the Royal Greenwich Observatory and McGill College Observatory, and in August 1891 the astronomer royal sent the necessary instruments to Montreal for the two Canadian stations, Canso, Nova Scotia, and Montreal, to provide time for familiarization. The other two stations involved in the campaign were Greenwich Observatory and Waterville (County Kerry) Ireland. The four observers were Professor C.H. McLeod of McGill, Otto J. Klotz, DLS, of the Department of the Interior, H.H. Turner, and H.P. Hollis, both of the Greenwich Observatory. The work was conducted in the summer of 1892 in four stages as follows:

	Montreal	Canso	Waterville	Greenwich
1	Klotz	McLeod	Turner	Hollis
2	McLeod	Klotz	Hollis	Turner
3	Turner	Klotz	Hollis	McLeod
4	Klotz	Turner	McLeod	Hollis

Klotz records that his transit instrument was one of a series used in the transit of Venus in 1874 (not to be confused with the 1882 transit), and

18 Sessional Paper no. 63, Department of the Interior no. 15, Report of Otto Klotz, 6 Feb. 1893

that the other observers had similar ones. The observer took his individual transit and level with him from station to station and mounted it on the fixed pier. The Klotz instrument had a clear aperture of $2^{31}/_{32}$ inches, a focal length of 36 inches, and setting circles 3 inches in diameter reading to minutes. Klotz further explains that a night's program involved four sets, LE, LW, LW, and LE respectively (LE and LW refer to the position of the striding level which is reversed from east to west when the transit instrument is lifted clear from its pivots and returned in the reversed position). Each set included one polar, one sub-polar (if available), and five or six other stars distributed between the zenith and 20° south declination. The stars were selected from the *Berliner Jahrbuch*, plus eleven additional ones from the *Nautical Almanac*. Each night, regardless of the weather, timing pulses were exchanged across the network.

The definitive result of the 1892 campaign was computed after much effort and was announced by the astronomer royal in 1896 when he visited Montreal on his way to Japan. The longitude of Montreal (centre line of the transit instrument) was $4^h 54^m 18^s.670$. In comparison, the campaign of 1883, based on the longitude of Cambridge, combined with the subsequent survey work of C.H. Sinclair of the Coast and Geodetic Survey, the former value for Montreal was $4^h 54^m 18^s.565$, so that the new value was greater by $0^s.105$.[19] In the 1897 report of the US Coast and Geodetic Survey, the Montreal value was included in the adjustment of the longitude net of the United States by C.A. Schott. The resultant value for Montreal was then adjusted to $5^h 54^m 18^s.634$, and it formed the base to which Ottawa was attached.

The federal decision of 1868 to establish a government observatory and time ball in Halifax had never been implemented. Instead, as we have seen, the timekeeping interests of the harbour were being served by a successful businessman, Robert H. Cogswell. In 1890 the government marine agent in Halifax wrote to his superior in Ottawa:

I am not aware of any arrangements having been made here for the giving of time by the dropping of a ball at noon. Since the year 1876 Mr Cogswell has been paid $100 per annum [from Ottawa] for furnishing to Citadel Hill here the correct time at 12 o'clock noon daily for the gun fired at that hour by the military authorities. The time is ascertained by Mr Cogswell by transit observations and is transmitted from his place in town by means of the telegraph (private line) to the Citadel Hill and the gun is fired accordingly.

19 Report of McGill College Observatory to the minister of marine and fisheries, 11 March 1897, Sessional Papers, Marine and Fisheries, 1897, Appendix C

The time given by the gun is nearly correct, but not sufficiently so for the purpose of rating chronometers. Almost all vessels here send their instruments to Mr Cogswell's business place to be rated.

By relying on Cogswell Halifax thus used an independent source of time even after the telegraph made central Canadian time available. During 1871 when the Department of Marine and Fisheries was developing a system of weather reporting stations, one of these was established in Halifax. It was a telegraphic reporting station, and the direct connection with Toronto made it convenient for the Halifax weather office to receive a time check whenever Saint John had an exchange of clock signals. This service was only discontinued in April 1896. Special time checks were also provided from Toronto for the benefit of the British survey gunboat *Rambler* at Halifax in 1898. But this service too was superseded by the time service from Saint John, which provided a signal automatically each day at 10 A.M. to the telegraph network. Special signals were also sent by request during the year 1900 from Saint John to the Royal Navy at Halifax, to North Sydney and Halifax for the British and French cable ships, to Mr W. Bell Dawson for the use of the Tidal Survey, and to others. At no time, however, was there mention of Mr Cogswell's time service being drawn into the federal service under the general supervision of Toronto.

Instead plans were carried out for a time ball in Halifax to be operated by the Saint John observatory. The Halifax press followed its installation with interest. A report in the *Nova Scotian* in March 1904 stated that, a short time before, an official from the meteorological office in Saint John had visited Halifax and located the site for the time ball, pending completion of the Custom Building. After viewing the city from the waterfront, the official chose the Citadel as the best location.

On 5 August 1904, the *Nova Scotian* reported that the director of the Meteorological Observatory at Saint John was in Halifax to superintend the final arrangements for the installation of the time ball, and to make the necessary electrical connections. The ball, to be located on the Citadel, was to be operated from the meteorological office in the Custom House in Saint John. That year 1 October marked the inauguration of the time ball service in Halifax. Its operation was described a few months later by the head of the Meteorological Service:

The ball is on a staff with base and small house for protection of the hoisting gear and electric release [Plate 4]. It is situated on the citadel a little north of the main signal station ... A clock especially designed for this (release mechanism) service was placed in the Western Union Office at Halifax. This clock has a good movement and a mercury pendulum, is wound electrically and is daily corrected or synchronized by the final dot at 10 A.M. of the signal sent by our transmitting clock ... The ball is automatically and electrically dropped at the instant of 1 P.M., the times of hoisting half elevation, full elevation and drop being synchronous with the ball at Saint John. The master clock in Halifax sends a signal to the citadel every hour and corrects a subsidiary clock placed there by the Meteorological Services for the guidance of the hoisting man. The Royal Engineers have charge of the hoisting mechanism and Mr C.W. McKee, manager of the Western Union has charge of the electric clock. To keep check on the time of the Halifax clock it is fitted with a break circuit attachment which registers a signal on the chronograph in Saint John together with one of our standard clocks. It shows only a small rate during the day. No failure to synchronize it has occurred. On two days when wire trouble existed the Halifax ball was dropped from Saint John, the same signal which synchronized their clock dropping the ball.[20]

Saint John had become the time and weather reporting centre of the maritimes, distributing daily by telegraph not only the time but also, after 1895, weather forecasts from Toronto covering the coast from Chatham to Boston. The observatory continued to improve its facilities for astronomical time determination. In 1895 a new Troughton and Simms transit arrived from London, having an aperture of $2^{1}/_{2}$ inches and a focal length of about 30 inches, and shortly thereafter electric lights were installed. Yet in spite of technical improvements the time exchanges indicated that the Saint John and Quebec clocks differed from the reference by more than a second on several occasions, the reference clock being the Standard Observer, the mean of the Toronto and McGill time determinations. In 1899 a clock by Victor Kullberg of London was added, and in 1901 the observatory was equipped to transmit by telegraph an automatic time signal daily from 8:58 to 9:00 A.M. 75th meridian time, which was the same as 9:58 to 10:00 A.M. Atlantic time. It became known as the ten o'clock signal, and was soon recognized as the official time throughout the Maritimes.

During 1902–3 the Saint John observatory was re-equipped, a vote of confidence in its ability as an independent source of time. The first item received was a drum chronograph. The drum, seven inches in diameter, could be driven at either 1 rpm or 2 rpm, and at the slower speed would record for $2^{1}/_{2}$ hours. Now the passage of the star behind successive lines

---

20 Report of D.L. Hutchinson, director of the Saint John Observatory, to R.F. Stupart, Oct. 1905, Sessional Paper no. 21, Department of Marine and Fisheries, A 1906

of the reticule could be recorded by pressing a key and compared to the clock beats which would also be recorded. Then came a new and larger transit, by Troughton and Simms of London. It was capable of higher precision than the older type: 'The instrument has a reversing carriage, and with the delicate level attached to one of the 6-inch finding circles and micrometer which is available in declination as well as right ascension, may be used as a zenith telescope as well as a transit. Small electric lamps are used for the illumination.'[21] The observing pier had to have its top section rebuilt to take the larger instrument. Finally, the observatory acquired sidereal clock No. 94 by Dr S. Riefler of Munich. The finest design of clock then available, it was enclosed in a glass cylinder which could be made airtight and maintained at a constant pressure. A mercury barometer was enclosed in the case. The riefler free escapement, combined with the nickel steel pendulum, made it a superb timekeeper. The clock, the new transit, and the drum chronograph made the Saint John time room fully modern.

In the McGill observatory, too, equipment was steadily improved, on both the meteorological and timekeeping sides, although the director, C.H. McLeod, had to struggle with the chronic shortage of funds common in universities at the time. In his annual report for 1894, McLeod made a strong plea to the minister of marine and fisheries for improved weather recording and forecasting equipment to enable him to fulfil 'a very persistent and increasing demand on the part of the public of Montreal and vicinity for special weather forecasts.' And he had the following to say about the time service:

Determination of clock errors have been made by the observation of 756 star transits on 134 nights. A determination of the clock errors is made in the following manner: – A comparison of the sidereal clock, and the mean time clock is obtained on the chronograph. The transits of six stars (one polar star and two equatorial stars, in each of the reverse positions of the instruments) are then observed and recorded on the chronograph. The inclination of the axis is measured before and after the observations of the stars in each position. The observations being completed, the clocks are again compared. The chronograph sheet is then read and the observations recorded, the instrumental errors deduced, and finally the clock errors are obtained. The error of the sidereal clock is allowed to accumulate, whereas the marking of the mean time clock is made to correspond to the local

21 Report of D.L. Hutchinson, Saint John Observatory, to R.F. Stupart, 31 Dec. 1903, Sessional Paper no. 21, Department of Marine and Fisheries, A 1904

mean time on the 75th meridian known as Eastern standard time. All the signals issuing from the observatory correspond with the marking of this clock.

The noon time-ball, for the use of shipping, has been dropped on every week day during the season of navigation. Special signals have also been transmitted daily to the Montreal fire alarm office for the noon stroke on the alarm bells.

By means of the automatic system of clock signals, which has been in use here for several years, a knowledge of standard time has been widely distributed through the corporations and institutions named below: –

The Canadian Pacific Railway Co., transmitting it daily to all stations along their lines to the Pacific coast.

The Grand Trunk Railway Co., through the great Northwestern Telegraph Company, for all their lines east of Kingston.

The Great Northwestern Telegraph Co., transmitting it daily to all the telegraph stations in eastern Ontario and the province of Quebec.

The Harbour Commissioners at Montreal.

The time signals of this observatory are also transmitted through the Great Northwestern Telegraph Company to Ottawa, for the firing of the noon gun at the Parliament buildings [Plate 16]. I regret again to have to state that the imperfect arrangements at Ottawa in connection with this service are such as to make the noon signal quite unreliable as a time standard for Ottawa ...

Exchanges of clock signals with the Toronto Observatory were made on 19 days. The average of the differences obtained between the mean time clocks of the observatories is 0.25 sec., and the greatest difference on any one day was 0.68 sec. The comparisons for the year show that the probable error of the time as given by one observatory at any time as compared with that given by the other, is 0.20 sec.[22]

The wide distribution of the McGill time service was described in August 1904 in the *Montreal Witness*:

The McGill Observatory clock supplies the correct time daily to all the stations on the Grand Trunk Railway system and to points on the Canadian Pacific and Intercolonial Railways from St John, NB to Bamfield, BC. The McGill time is automatically repeated at Canso, NS to the Azores Islands; the inhabitants of Bermuda and Jamaica receive it from Halifax, where it is repeated by hand; the British warships correct their chronometers (get it from CPR) by it at Halifax and Victoria, and the German fleet does the same at the Azores (get time by signals from the land); the operators at the islands in the Pacific where the Australian

---

22 Sessional Paper no. 11, Marine and Fisheries A 1895. C.H. McLeod to the minister of marine and fisheries, 31 Dec.

cable lands, set their watches by it, and it automatically regulates a large number of the electric clocks in the city.

The CPR has an official timekeeper at all its terminal points, who corrects the watches of the trainmen ... from the daily messages received from Montreal. This message is sent through at 11:54 A.M., and the seconds are ticked off until 11:56 A.M., when the message closes. During the first minute, a single 'click' identifies each second, and during the second minute a double click is heard. Automatic repeaters at Fort William, Winnipeg and Swift Current take the signal to the western terminus.

About 3/100 of a second is occupied in passing through each repeater, and the time occupied on the wire itself is about 2/100 of a second. Thus the actual time consumed between Montreal and Victoria is about 15/100 of a second.

While a certain degree of civic pride may have been exaggerating what McGill time signals were accomplishing unaided by other sources, this account does draw attention to a unique aspect of Montreal time. Whereas Toronto time spread its influence eastward, so to speak, to Quebec and the Maritimes, Montreal's moved westward across the country, through the telegraph lines of the Canadian Pacific Railway, co-ordinating train schedules and serving communities along the way. The importance of early information on westerly weather patterns led to the establishment in July 1890 at the naval base of Esquimalt of the first meteorological telegraphic reporting station in the province, under the charge of E. Baynes Reed. Although he was equipped with a chronometer, which he checked daily against the McGill time signal, there is no indication that Reed made any contribution to timekeeping in Victoria, or to the firing of the gun at noon and 9 P.M.

The earliest inhabitants of that coastal community had depended for time checks on selected ships, commercial and naval, which would fire a gun at an appointed hour while in port. Later, a land-based time gun came into use by and for personnel of the military base at Esquimalt. In May 1877 the Victoria *Colonist* printed a circular letter sent out by Admiral Horsey, commander of the British Pacific fleet:

It may be of interest to the Captains of ships and others who may wish to rate their chronometers, or to correct their watches, to know that when one of HBM [Her Britannic Majesty's] ships is present in any port in the Pacific, such ship (or the senior, if more than one) marks the time daily by a time ball at noon.

The ball is hoisted 'at the dip' at about 11:55, is hoisted right up at about 11:58, and is dropped at the instant of noon mean time.

When the British Flag Ship is present, some persons are in the habit of correcting their watches by the evening gun; but it should be understood that the gun only marks the time approximately, whilst in the case of the time ball, the limit of error should not exceed half a second.

Lacking a military post to take responsibility for signals, Vancouver experienced more difficulty with its early timekeeping. In 1894, a 'nine o'clock gun' took over the task from the bell atop the old Water Street firehall. In 1898, this was replaced by a time signal on Deadman's Island:

It was decided to fire a dynamite cartridge hoisted at the end of a jib and connected by wire with the CPR telegraph office in Vancouver each day at noon, but it was subsequently found that the noise of the city drowned the sound and therefore it has since been fired at 9 P.M. The cartridge is prepared and placed in position for firing by Wm D. Jones, keeper of the bell tower at Brockton Point, and at the proper instant an electric contact is made at the telegraph office by the chief operator who rates a chronometer, provided by this service, by the time signal given each morning over direct wire from McGill University, Montreal, by Professor C.H. McLeod. The accuracy of the signal is therefore dependent on three things: firstly the accuracy of the time as given from Montreal; secondly the uniform rate of a chronometer during twelve hours from 9 A.M. to 9 P.M.; and thirdly, the trustworthiness of the operator at Vancouver. It is proposed very shortly to install a gun in place of the dynamite cartridge, as the fire of the gun will probably be heard more generally.[23]

The following year the gun was restored [Plate 13]. But difficulties persisted, and in 1902 Baynes Reed went over from Victoria to investigate and found another source of error, the wire from the telegraph office to the gun. The gun itself was in danger of tumbling down the embankment into the water because the shelter was falling apart and the foundations of its mount had crumbled. Mr Brown, the local meteorological observer in Vancouver, whose services were carried out with great care, was not involved with the gun, which was the combined responsibility of the gun keeper and the telegraph operator. It is possible that this separation of duties may have been at the root of the difficulties.

A new generation of observers and administrators was coming forward. In Toronto, Charles Carpmael had been succeeded in 1894 by R.F.

23 Sessional Paper no. 11, A 1900, Meteorological Service, Department of Marine and Fisheries

Stupart as Dominion meteorologist. In 1898 the dynamic young F. Napier Denison [Plate 12] had been posted to Victoria, where the service had moved from Esquimalt, to assist Baynes Reed, particularly in the operation of a new Milne seismograph for measuring tremors in the earth's crust. In 1906, Brown, the Vancouver observer, retired and was replaced by T.S.H. Shearman, an amateur astronomer who understood the work involved [Plate 15]. Denison in Victoria and Shearman in Vancouver each sought to establish an observatory as an independent source of time. In 1911 Shearman wrote to Stupart to announce that 'plans have been discussed here to establish an astronomical observatory in this vicinity for the purpose of distributing time by the wireless method to the rapidly increasing shipping of this coast and for the rating of chronometers, the study of solar physics, terrestrial magnetism, and in an auxiliary observatory, to use the clean air of the mountains to the north of the city, and in that way to do our share in the establishment of the chain of international observatories strung like pearls along the summits of the Rockies and the Andes Mountains.' It would be of great advantage to have the proposed observatory connected with the British Columbia University soon to be erected at Point Grey. A resolution to this effect was sent by the Board of Trade to the Hon. Dr H.E. Young, provincial secretary in Victoria.

Shearman referred to a remark in the Victoria paper to the effect that the proposed observatory for Vancouver was merely an attempt to take something from the city. Not so, Shearman insisted; he had for the past twenty years advocated an observatory for British Columbia, and now his plans called for a ten-foot reflector for Grouse Mountain. Denison, he said, was simply repeating Shearman's work and claiming it as his own: 'Denison is welcome to any improvements to his new seismological observatory, but he is not, as the Victoria papers state, the only man in Canada capable of conducting a BC seismological observatory.'[24]

Denison, inclined to act first and ask later, had in 1909 purchased with his own money a small transit telescope (five years later the Department agreed to reimburse him the $200), and in 1912 he acquired a telescope that had been made by a local amateur astronomer, O.C. Hastings.

The space provided in the Victoria Post Office for the Meteorological Service was becoming inadequate, and Denison submitted to Toronto a well-devised plan that would provide facilities for his astronomical work

24 T.S.H. Shearman to R.F. Stupart, 6 June 1911, Meteorological Branch correspondence file, Public Archives of Canada

as well as an excellent site for a seismograph.[25] It would be an observatory at Victoria on Gonzales Heights adjacent to the government radio station. Out of this plan grew the Gonzales Observatory, overlooking Victoria and the Strait of Juan de Fuca, which finally came into operation in April 1914 [Plate 14].

The previous December Denison had outlined the requirements for his time service, which included a good sidereal pendulum, a good mean time pendulum, a chronometer, and an electric contact, either hand-operated or attached by micrometer drive to the telescope. He also submitted plans with estimates of cost for a time ball similar to the one at Halifax. Stupart had some misgivings about Denison operating a time service. In March 1915 he wrote, 'Before proceeding further with this matter I want your definite assurance that you are sufficiently conversant with the adjustments and use of a transit instrument to carry on the time work in a way that will be a credit to the [Meteorological] Service. I ask this because I have no knowledge of your ever having done such work, and it is altogether important that when once begun, the service shall be continued in a creditable manner.'[26] Denison replied: 'I beg definitely to assure you that I am now quite conversant with the adjustment and use of the transit instrument, and to insure most accurate results, Mr W.S. Drewry, a prominent Provincial and Dominion Surveyor ... is kindly assisting ... in those parts of the observational and reduction work I had not perfected when learning under Mr F.L. Blake at Toronto some years ago.'

By now, because of failing health, Baynes Reed was absent from the office, and Denison shouldered the responsibility for all the work, in which he was so absorbed that as a measure of economy he and his wife took up residence in the observatory itself. For the next twenty years this modest one-room apartment served as their home.

Denison pushed forward plans for a time ball, which was successfully mounted on the Belmont Building and went into operation in May 1915. In a subsequent report he described its operation:

The time for this service is obtained from star observations taken nearly every clear evening by means of the transit telescope, which is well mounted in a special

25 F.N. Denison to R.F. Stupart, 10 Dec. 1912, Meteorlogical Branch correspondence file, Public Archives of Canada

26 R.F. Stupart to F.N. Denison, 18 March 1915. Meteorological Branch correspondence file, Public Archives of Canada

room upon a massive concrete pier which rises from the solid rock. The adjustments for collimation and azimuth have been carefully made, and a fine fixed mark for checking these has been established on a government building at a distance of two miles ... It is possible to keep the time accurate to within one or two tenths of a second.

The time ball, which is a distance from the observatory, about two miles, is controlled by a telegraph key here in the following manner: At 12:30 P.M. each day (Sunday included) the CPR Telegraph Co. connects our line with the time ball circuit; at 12:45 P.M. a signal of two taps given from here notifies the man in charge of the time ball to hoist it to half mast; at 12:55 P.M., three taps notifies the man to hoist the ball to the top and to set the electric trigger. At one half second to the exact 1 P.M., I press the key here and the ball drops on true time. The operation is checked daily by watching the ball through a telescope here.[27]

Meanwhile, on the mainland, the busy and expanding port of Vancouver brought an increasing demand for an adequate time service for marine chronometers. Shearman's requests were supported by Stupart. In 1913 the latter reported as follows: 'At Vancouver I arranged for a "Time Service" under the supervision of Mr T.S.H. Shearman, exactly in accord with the recommendation made with my estimates furnished the Department under date of May 29, 1912. Mr Shearman will have a transit, and his observing station will be connected by wire with a room in the New Public Building to which the Masters of ships may bring their chronometers for comparison and rating and where the public may apply for information regarding climatic conditions of the Province ... Mr H.H. Stevens, MP ... is quite satisfied with the arrangements I have made, and personally inspected the new Meteorological room.'[28] However, there is no evidence that a transit instrument was ever put into service by Shearman.

Shearman attempted to secure a time ball and storm signal mast for Vancouver, with full support from Stupart. The plan failed when in February 1917 the Public Works Department was obliged to refuse him the use of the Post Office roof. But Shearman's office became recognized as the source of correct time where ships chronometers could be brought for rating, and from which the signal was given for the firing of the 9 P.M.

27 Sessional Paper no. 21, A 1917, Meteorological Report, Department of Marine and Fisheries.
28 R.F. Stupart to deputy minister, Department of Marine and Fisheries, Meteorological Branch correspondence file, Public Archives of Canada

gun. It was a telegraph operator, though, and not Shearman, who came in to operate the key which fired the gun, a fact giving rise to disputes over accuracy. For the time being, then, Shearman was left dependent upon the daily time signal relayed from the McGill Observatory, while Denison forged ahead with rather inadequate equipment to maintain an independent source of fundamental time.

Back east the accuracy of time determination was slowly improving through the use of better instruments. In Toronto, the expansion of the city's street railway and of the attendant wires carrying heavy electric currents was beginning to have a serious effect on the magnetic records obtained at the observatory. So in June 1898 a new observatory building was commenced on a quiet site at the village of Agincourt, about ten miles to the northeast. Stupart devised a break circuit for the mean time pendulum so that time pulses could be sent automatically for the weekly time check at Agincourt. The daily signal which rang the Toronto fire alarm bells at 11:55 A.M. and the daily signal to the CN telegraph were also controlled automatically.

On the occasion of the Geophysical Conference held in Toronto on 8 September 1904 a direct telegraphic time comparison was made at midnight with Washington. The difference in time was five hundredths of a second. On 6 May following, another comparison was made with Washington which showed exact coincidence.[29] Communication between these two cities was a natural consequence of the fact that they were both data centres for meteorological research, and custodians of correct time.

Transit observations for the fundamental determination of time at Toronto continued year by year, the number of nights varying from a low of 57 in 1905 to a high of 139 in 1925. During the earlier years an occasional transit observation of the sun was made to supplement stellar observations during periods of poor weather. After the first world war no further solar observations for time were reported, and in 1923 the transit work was turned over to the Department of Magnetism under W.E.W. Jackson. By now the work at the central office in Toronto had grown to include weather forecasting, atmospheric physics, climatology and agricultural meteorology, terrestrial magnetism, and astronomy, the last comprising the maintenance of the time service and solar observations for sunspot numbers.

Correct time was important throughout the Meteorological Service, whether the observations were made at the well-equipped principal re-

29 Sessional Paper no. 11, 1906, Meteorological Report, Marine and Fisheries

porting centres such as Montreal, Quebec, and Saint John, or by any of the large number of secondary posts, manned for the most part by capable volunteers. In the maritime provinces the daily signal from Saint John was available wherever the telegraph went. The rest of Canada was served by the McGill signal, which was also distributed over the telegraph network. Stupart and his staff made inspection trips each year attempting to visit each station at least once every two years, arranging to have equipment repaired or replaced, and encouraging the operators in their work.

Yet improvements in accuracy were by now more of scientific than of public importance. In 1918 James Weir, C.H. McLeod's successor in charge of the McGill Observatory, remarked about his time service that 'the probable error of the determinations [for time] is several hundredths of a second. Greater precision is futile in view of the timekeeping possibilities of the astronomical clocks in use, and the fact that the time signals given out, if kept well within the second, are subject to no criticism from the city, shipping and railway services.'[30] Of more importance to the public was accessibility to correct time. One way in which McGill time was being made more available was mentioned by Weir: in addition to the country-wide distribution of the time signals, 'several jewellers requiring an accurate time standard have installed loops and ticker service. Rental of the lines is chargeable against this service and the electrical work is attended to by the Dominion Gresham Company.'

A detailed report for 1923–4 indicated the extent to which the Saint John time service had permeated the Maritimes.

The transmitting clock is run on Atlantic Standard time and has a very small daily rate. After comparison with the Sidereal clocks any outstanding error is adjusted by a switch outside of the protecting case that electrically controls two small weights which by this means may be placed on or off a small shelf attached to the pendulum, it may be accelerated or retarded and usually in a few minutes exactly corrected. The well known code of signals from this clock are entirely automatic.

Three loop lines connect the observatory with the outside time signal service, two of these loops run to the Telegraph Office and one to the Telephone Office.

One loop line from the observatory to the Western Union Office is also extended to the time ball tower on the Customs Building here and is used at 1 P.M. for automatically releasing the time ball at that hour. The clock signals are widely

30 Sessional Paper no. 21, A 1919, Meteorological Report, Department of Marine and Fisheries

disseminated throughout the Maritime Provinces and are also received at all telegraph offices on the Intercolonial division of the Canadian National Railways and their branch lines, as well as at the Dominion Atlantic Railway in Nova Scotia.

At many points, such as Halifax, Truro, the Sydneys, Moncton, Charlottetown, etc., the Telegraph Company has installed clocks which are electrically wound and daily corrected by the 10 A.M. signal; this is done by one of their operators throwing a switch during the ten seconds pause before the dot made at the hour exactly and throwing it open during the ten seconds pause of safety after the hour. These clocks have second hands and are set to the second. As our time is available to the Telegraph Company every hour they may correct any of these clocks at hours other than 10 A.M. These corrected clocks being installed in the Telegraph Company's public offices, the public are afforded the opportunity of obtaining the correct time.

Another contact maker on the transmitting clock which closes the circuit on the 59th second and opens it at the hour exactly is purely for local purposes and is used for the correction of tower, hotel, street, bank, factory and watch and chronometer rater clocks. [Rater clocks were usually found in a watchmaker's shop.] In these cases the clocks are purchased outright and the signal of correction automatically is sent from our transmitting clock every hour day and night on a loop connecting the observatory clock relay with a special switchboard in the operating room of the Telephone Company, who charge an annual rental for the wire service to the owners of the clocks. Within a prescribed radius the fee paid to the Telephone Company by the users is ten dollars yearly. Some watchmakers have bells or sounders which give them a signal stroke every hour. Another loop in the Telegraph office is used for synchronizing clocks in their Saint John Offices and operating room.[31]

By the 1920s, however, a new technology, radio, was revolutionizing time distribution as profoundly as the telegraph had done seventy years earlier. With radio, electrical simultaneity over great distances became possible without the intervention of human or mechanical relays or the necessity of complex and expensive wire connections. The scientific utility of radio in timekeeping was shown during the transit of Mercury in May 1924, when L.B. Stewart, professor of engineering at the University of Toronto, assisted in setting up a radio receiver so that the Toronto sidereal clock could be compared directly with the Washington time signals. Two years later, in connection with the World Longitude Campaign, observers at Brockton Point in Vancouver used a modern receiver

31 Sessional Paper no. 28, A 1925, Report of the deputy minister of Marine and Fisheries

to secure time signals from Washington, England, and Paris. The universal availability through radio of time signals from highly equipped observatories made astronomical time clocks less important in centres that chose to specialize in other scientific endeavours. In 1919 the McGill observatory, for example, acquired a wireless receiving set to monitor Washington time signals emanating from Arlington, Virginia. Thereafter, transit observations apparently declined at McGill; the annual report of 1924 stated that they were made on only sixty-seven nights. Two years later, in 1926, the transit hut was demolished to make way for new construction and was never re-established. After nearly seven decades, McGill's contribution to the determination of astronomical time came to a close.

On the west coast, both Shearman and Denison recognized the shortcomings of the telegraph time signal from Montreal. Shearman requested that he be supplied with a radio capable of picking up the USA time signal daily from Mare Island, California. In support of this request, Stupart remarked in 1921 that 'the time received by wire in Vancouver is from Montreal, not Victoria, and is not received automatically from the McGill Observatory clock. The Observatory clock records in the main Montreal Telegraph Office, but the signals are transmitted over the Western Telegraph wires by hand, and are found to be not sufficiently accurate for purposes of rating ship chronometers.' 'I place no credence' he went on, 'in Mr T.S.H. Shearman's statement that the Gonzales Time signal is unreliable, except to the extent that as Mr. Denison's equipment is inferior to that at Mare Island, California, his observations are not capable of the same degree of refinement. I am sure however, that the observations are so carefully taken that an error of a second would be rare. I think, however, on the whole it would be well to install a valve receiving equipment which will enable the Vancouver observer to pick up the San Francisco signal.'[32] Denison in response proposed that in place of a $600 radio for Vancouver, a $200 crystal set be purchased, and the remaining $400 be used to acquire for Victoria a modern observatory sidereal clock with electric attachment and a good chronograph. He would then be able to guarantee daily wireless signals from Victoria with one-tenth of a second in accuracy. Neither proposal was accepted.

Nevertheless, within two years Denison, who found the transit work difficult with his equipment, was listening to the Mare Island radio time

32 R.F. Stupart to the deputy minister, Department of Marine and Fisheries, 30 Nov. 1931, Meteorological Branch correspondence, Public Archives of Canada

signals as relayed over the telephone from a receiver of the Department of Marine's Radio Service next door. It was an awkward and at times inconvenient arrangement, so E.J. Haughton erected an antenna and assembled a valve receiver for Denison at a total cost of $90. With this receiver both Washington and Mare Island signals were monitored, and Gonzales ceased to be an independent source of time. Haughton then wrote to his superior, C.P. Edwards, and Denison reported the expenditure to Stupart. Since it was all within the same Department of Marine and Fisheries, the deputy minister simply made a book entry against Stupart's account.

Transit observations gradually ceased in Quebec later in the decade, but they were maintained in Toronto until February 1931, when a long-wave receiving set was installed for the purpose of checking time with the Arlington signals. The result of daily checks through the month of March gave Toronto–Arlington = −0.07. Time signals were telegraphed once a week to the magnetic observatory in Agincourt to check and control errors of the clocks and chronometers.

The exception was Saint John. After 1931 the fundamental determination of time had ceased to be performed at any of the other meteorological centres across Canada. Instead reliance was placed on the time which originated at either Ottawa or Washington. But in Saint John, because of public pressure, the transit instrument continued in use on clear nights, and the network of time circuits controlled by the observatory remained active until F.M. Barnes, the director, reached retirement age in 1949. The radio era revolutionized the distribution of time as well. Perhaps the first to recognize this fact was Napier Denison in Victoria. In 1917 in co-operation with E.J. Haughton he had commenced a daily 10 A.M. wireless time signal by hand key from the observatory through the Gonzales Heights wireless station. The signal was reported to have a coverage of two to three hundred miles and to be of benefit to shipping and to other radio stations. An automatic sender was later designed to operate from one of the pendulum clocks, which provided uniformity within a few hundredths of a second for the two minutes of radio transmission. Commencing in 1923 the wireless time signal from Gonzales Heights was relayed by hand from the powerful station at Estevan Point, thereby increasing the radius of coverage to three thousand miles. And a 10 P.M. radio time signal was started.[33]

33 Sessional Paper no. 48, 1925, Department of Marine and Fisheries, Report of the
    deputy minister, 67, Public Archives of Canada

At the other end of the country a radio transmitter was built at Camperdown, on a height of land seaward from Halifax harbour, which continued as a radio communication point till a new station replaced it in the late 1960s. The time signals were sent by land line from Saint John to the transmitter (vcs, 417 kcs), then relayed by hand. Foreshadowing the future, when radio receivers spread through private homes, was the report from Saint John that 'Time signals have since December 1924 been broadcasted from CNR radio station at Moncton (CNRA) on Tuesdays and Fridays at 10 P.M.'

The ten o'clock signal, distributed from the Saint John observatory throughout the Maritimes over the CPR and Western Union telegraph lines, was also broadcast from the Marine and Fisheries coastal radio station at Chebucto Head, adjacent to Halifax. At Moncton, the time signals were broadcast at 3 P.M. each week day, and at 10 P.M. on Tuesdays and Fridays. In October 1926 the director of the Saint John observatory described the nature of the time signal: 'Beginning at 9:58 A.M. Atlantic time, dots are made each second up to and including $9^h\ 58^m\ 57^s$, then a pause of two seconds, followed by a dot at $9^h\ 59^m\ 00^s$. Then a pause of two seconds follows. The clock then makes dots each second up to and including $9^h\ 59^m\ 50^s$, then a pause, followed by a dot at 10 A.M. Atlantic, or standard time of the 60th meridian west longitude, equivalent to 2^h P.M. GMT.'

The same code was built into the sending mechanism used by Denison in his broadcasts over the Marine and Fisheries transmitters at Gonzales Hill and at Estevan, BC.

An official statement concerning radio time signals broadcast by the Department of Marine and Fisheries was sent, in 1926, by the director of radio, C.P. Edwards, to R.M. Stewart, director of the Dominion Observatory:

Station	Call	Wavelength	Character of transmission
Chebucto Head, NS	VAV	600 Metres	Synchronous spark, 480
Estevan, BC	VAE	600 Metres	Synchronous spark, 480
Gonzales Hill, BC	VAK	600 Metres	Valve ICW

The most visible result of radio communication was the disappearance of time balls in the late 1920s and early 1930s. With time signals on their radios, ships had no need of such approximate visual checks for their chronometers. And yet, like time guns, a time ball could be kept functioning on public demand. When the Saint John time ball was out of commission for two months in 1934, owing to a breakdown in the hoisting and dropping mechanism, public concern caused it to be restored, and it remained in service for another decade or more.

The fact that clocks could now be checked by radio signals instead of by transit observations spelled the doom of local time rooms. By 1930, for instance, the Quebec observatory, in its timekeeping capacity, did no more than relay correct time, first from McGill, later from Ottawa. But other centres did not vanish so quickly. The most vigorously independent, as we have seen, was Saint John. In 1936 it was still setting the clocks of the Maritimes:

Transit observations with the three-inch meridian telescope for the determination of time have been made as frequently as possible and comparisons made with the Riefler and Kullberg sidereal clocks. The Riefler sidereal clock situated in the basement of the observatory continues to give excellent service.

Time signals are automatically sent from our mean time transmitting clock every week day over the Canadian National wires to all their offices in the Maritime Provinces as well as to the Dominion Atlantic and Canadian Pacific Railway Companies. Automatic signals for broadcasting are sent over the Canadian Pacific land lines to Camperdown.

The daily time signals in Saint John have been given to shipping and others by the dropping of the time ball on the Customs House. In Saint John the system of hourly synchronizing of office, street and tower clocks continues to give most satisfactory and useful service. The comparison of time between the telechron clocks throughout the city and the mean time transmitting clock continues to give satisfaction. The number of telephone calls for correct time has increased.

The equipment has been maintained in good condition throughout the year. Minor repairs were made to the impulse clock and the tower clock.

Many visitors, including Junior service clubs were welcomed to the observatory during the year and conducted through the building.[34]

In spite of the fact that the responsibility for correct time was officially transferred to the Dominion Observatory in Ottawa in 1936 the observatory in Saint John remained operative long afterward. The time ball on the Customs Building became inoperative in 1946 and was never restored to service. Fundamental timekeeping based on astronomical observations ceased as of May 1947, giving way to radio time signals from Ottawa,

34 Report of the assistant deputy minister, Department of Marine and Fisheries, for the fiscal year 1935–6, 52, Public Archives of Canada

London, and Washington. And even in closing the observatory in 1951, C.S. Beals, the Dominion astronomer, found various public and private groups arguing that they were being deprived of a service that was rightfully theirs.

The McGill signal continued to serve the widest area of the country. The acquisition of a radio receiver for Washington time in 1919 had been intended to improve the time room, not replace it. A second Howard astronomical clock was simultaneously acquired from the Geodetic Laboratory, raising the complement of precise pendulums to four: two sidereal and two mean-time. The two main activities which had been developed by C.H. McLeod, weather observing and time service, continued under the impetus he had given them, but with changing emphasis. When the Bell Telephone Company in 1922 discontinued its practice of answering requests for time, another load was added to the observatory. In the 1930s McGill provided a signal each day to the telegraph companies which was relayed Dominion-wide: except for Victoria, Toronto, Ottawa, and the Maritimes, the time for Canada was set by McGill. The Dominion Observatory time signal replaced the McGill signal for telegraph companies only in 1947.

The Toronto observatory continued to improve its time room until the mid 1930s. In 1931 it was reported that the time service was maintained with increased efficiency during the past year. Thermostatic control of the temperature in the clock room resulted in greatly improved clock rates and consequent saving in time necessary for transit work. Time exchanges were made about once a month with Quebec, Montreal, and Saint John and the results were quite satisfactory. The mean differences were Toronto–Quebec = −0.12 sec, Toronto–Montreal = −0.17 sec, Toronto–Saint John = −0.70 sec.'

For decades these observatories had provided correct time as well as weather information. And they continued to maintain a time service long after their transit rooms had ceased to function, their own clocks being corrected by radio time signals, all except Saint John, where the transit room remained active until 1947. The transit operations had ceased at Montreal by 1920, at Quebec by 1930, and at Toronto by 1931. The meteorological office in Toronto ended its supervision of timekeeping with the reorganization of December 1936 when timekeeping for Canada became the responsibility of the Dominion Observatory at Ottawa, established in 1905, which had developed an expertise in timekeeping, with modern equipment in both transit instruments and precision clocks. The magnetic and seismic survey activities of the Toronto observatory were also transferred to Ottawa, so that the Meterological Service was left free 'to devote all its efforts to provide a weather service for the Dominion.'

In the West, Sherman's contribution in Vancouver gradually declined, becoming primarily concerned with acting chronometers; his office was much better known for its weather forecasts. Denison in Victoria retired in 1935; after that date the observatory continued to provide a time service by receiving radio signals. But transmission of time from the observatory had ceased to be up to continually advancing standards, and it was discontinued in 1940.

The rise in importance of the Dominion Observatory in Ottawa had been shown in the 1920s by a difference of opinion between R.F. Stupart, Dominion meteorologist, based in Toronto, and Otto Klotz, Dominion astronomer, in Ottawa. It concerned the national implications of the new radio time signals. Stupart favoured the principle of receiving wireless time signals. He had recommended the acquisition of a radio receiver on the west coast to monitor the signals from San Francisco. Klotz, when asked for an opinion, voiced strong opposition to the proposal. He felt that the time service provided by the Dominion Observatory was superior to that which emanated from San Francisco. The acquisition of a radio was an unnecessary expense, he believed, particularly when the ultimate product was inferior to what was available at home. Moreover, he felt it was humiliating that any part of Canada should admit dependence on a foreign product when telegraphic communication could make the native product available with a reliability superior to that of radio.[35]

Undoubtedly there was some degree of rivalry between the two men. Both of them had developed a time service, the separate tentacles of which were commencing to overlap. And both had an active interest in the geophysical disciplines of magnetism and seismology. It was with some dismay that Toronto and McGill had seen a new observatory built in Ottawa at the turn of the century. Both institutions had urged the need for expansion and the continual up-dating of facilities in order to keep the routine operation of a time service attractive to fresh young scientific minds.

35 Memo from Otto Klotz to W.M. Cory, deputy minister of the interior, 25 Jan. 1922, was responded to by a memo from R.F. Stupart to the deputy minister, Department of Marine, 2 March 1922. Correspondence file, Meteorology Branch, Public Archives of Canada

3

The rise of the
Dominion Observatory

The story behind the building of an observatory at Ottawa in the early years of Confederation involves two historic events. One was the acquisition by the federal government in 1869 of the territory of Rupert's Land by the purchase of the rights from the Hudson's Bay Company; the other was the building of the Canadian Pacific Railway, in keeping with the promise to provide a link from British Columbia to the east in return for that colony's joining confederation.

With the purchase of the rights from the Hudson's Bay Company, there was made available for settlement in 1869 a vast area of rich fertile prairie land. 'It was realized,' wrote John S. Dennis jr in his *History of the Surveys Made Under the Dominion Lands System, 1869 to 1899*, 'that one of the first duties of the Federal Government was to devise and adopt a comprehensive scheme or system upon which to conduct the surveys of the country, and to proceed with the survey of such portions as were likely to be required for immediate settlement.'[1] The actual survey commenced in 1869 with Pembina, North Dakota, at the United States–Canadian border as the starting point. Telegraphically its longitude was determined with respect to Chicago, whose position in turn with respect to Greenwich had been determined previously, one measurement having been made from Canada, as we have noted, by E.D. Ashe of Quebec City in 1857.

From Pembina, a distance ten miles west was carefully measured and the first initial (or Winnipeg) meridian marked out, its position established as 97° 27′ 28″ west. From here successive initial meridians, designated as the 2nd, 3rd, 4th, 5th, and 6th initial meridians were identified at

1 John S. Dennis jr, 'A short history of the surveys made under the Dominion Lands System, 1869–89,'

102°, 106°, 110°, 114°, and 118°. In addition, the second initial meridian east was identified at 94°, so that guidelines for the grid extended across the western plains from Lake of the Woods to the Peace River area. The subdivision into large blocks, townships, and individual farm holdings required the services of a large number of surveyors for several years. Astronomical work in the field involved observations for latitude to check long distances measured with the chain. Theodolites of ten-inch aperture, capable of a resolution of five seconds of arc, were very effective in extending the survey. But checks in longitude were not attempted until the presence of telegraphic communication rendered precise determinations possible.

In British Columbia a strip of land twenty miles wide on each side of and parallel to the projected railway was given to the federal administration by the province. Because of the mountainous nature of the terrain it could not be surveyed by extending the lines of the prairie network. Instead it was necessary to run a detailed azimuth survey along the length of the line, and determine astronomically the latitude and longitude of a few positions so that the whole line could be related to the prairie network.

The success of this venture was largely due to the work of three men of about the same age, who rose to prominence as Canadian surveyors and who were instrumental in establishing an observatory at Ottawa. They were E.G. Deville, W.F. King, and O.J. Klotz [Plates 18 to 20].

Edouard Gaston Deville, born in France in 1849, was educated at the French Naval School at Brest and after six years of hydrographic work with the French Navy retired with the rank of captain and came to Canada in 1874. As a surveyor in the Province of Quebec in 1875 he assisted E.D. Ashe in determining the longitude of four places along the Ottawa River by telegraph connections with Quebec. Five years later, having qualified as a Provincial Land Surveyor (PLS), a Dominion Land Surveyor (DLS), and a Dominion Topographical Surveyor (DTS), he joined the western surveys. By reason of his experience and ability he was appointed in 1881 as inspector of surveys, in 1882 as chief inspector of surveys, and in 1885 as surveyor general on the retirement of Lindsay A. Russell. This office he held with distinction till 1922, when he was appointed director general of surveys.[2]

William Frederick King, born in 1854, came to Canada from England at the age of eight. A brilliant academic career was interrupted at the end of his third year at the University of Toronto when, in 1872, at the age of

2 J.D. Craig, 'Edouard Gaston Deville'

eighteen, he was appointed astronomical assistant to the boundary survey. At this time the 49th parallel between the Lake of the Woods and the Rocky Mountains was being identified. Two years later he returned to Toronto, and on graduation six months later was awarded the gold medal in mathematics. In 1876, he qualified as both DLS and DTS, being the first candidate to pass the DTS qualifying exam. In 1875 and 1876 he was astronomical assistant on the Special Survey of the Northwest Territories (that part of the Territories which later became Saskatchewan and Alberta), and from 1877 to 1881 was in charge of the astronomical section. The Special Survey was responsible for the initial meridians and base lines from which the remainder of the survey was extended. King collaborated with Deville in revising the *Manual of Surveys*, most of the tables being computed by King. In 1881, along with Deville, he was appointed inspector of surveys, and five years later became chief inspector. In 1890 he was appointed chief astronomer, an indication that astronomy was a recognized branch of the Department of the Interior. In 1905 he was appointed director of the Dominion Observatory, a post he held until his death in 1916, and in 1909 received the additional appointment of superintendent of the Geodetic Survey of Canada.[3]

Otto Julius Klotz, born in Preston, Ontario, in 1852, was a scholarship pupil in public and in high school. Mathematics, astronomy, and science were his main interests, which he studied at the University of Michigan at Ann Arbor, graduating in 1872. He practised for a few years as a private surveyor, qualifying in 1877 for his DTS. Two years later he was appointed a contract surveyor for the Canadian government, in which capacity he had several interesting assignments. He was engaged for a while in base line work, the highest class of surveys in the northwest, King being his immediate supervisor. In 1884, when the Hudson Bay Route was being examined, Klotz was commissioned to report on the land approach to York Factory. His adverse report, based on the shallow waters surrounding the proposed harbour, undoubtedly had some influence in turning official attention to Churchill as the rail terminus. Boundary problems engaged him along the north shore of Lake Erie and the coast of Labrador. But the assignment which had perhaps the greatest impact on the future of astronomy in Canada was the determination in 1885–6 by astronomical methods of the latitude and longitude of certain points along the Canadian Pacific Railway in British Columbia. He was in charge of the operation, with Thomas Drummond as his astronomical assistant

3 J.S. Plaskett, 'William Frederick King'

and William Ogilvie responsible for the azimuth survey. Klotz was pleased, in later years, to note that his classification as astronomer was the first occasion that that title had been given to an employee of the government of Canada. In 1892 he assisted in the transatlantic Greenwich-Montreal longitude determination, and a decade later extended the transpacific longitude determinations to Australia. He succeeded King as director of the Dominion Observatory at Ottawa in 1917.[4]

When Klotz commenced his work in British Columbia early in 1885 the railway was not quite completed, and the telegraph line had not yet been extended to the coast. The United States Coast and Geodetic Survey had extended the telegraphic network of longitude determinations to Seattle, so that Klotz used it as the base station from which to measure the longitudes of Victoria and Kamloops, and subsequently other locations in British Columbia. The telegraph, 'the lesser metallic girdle,' was completed just three days before the last spike was driven home on 7 November 1885. During the next two years, therefore, the telegraphic longitudes were extended eastward from Kamloops to Winnipeg and Port Arthur (Thunder Bay). And there is some evidence that a telegraphic exchange was made between Montreal and Winnipeg in 1888.[5]

Prior to his appointment as inspector of surveys in June 1881, King had been engaged as astronomical assistant in the verification of the position of those principal lines which were laid out as controls for the survey. He was in charge of the western part of the field work, and collaborated with the chief, A.L. Russell, based at Winnipeg. In spite of the frustrations with the less than perfect telegraph line, a total of twenty-one standard survey astronomical stations had been established by 1881.[6]

During the next three years, Deville moved to Ottawa, and King had the supervision of all the field work in western Canada. They were busy years, especially 1883, in part owing to the rapid progress of the CPR and the influx of settlers. By 1885 the activity had subsided. But from the standpoint of astronomy in Canada it was a most eventful year because the work along the British Columbia railway belt effectively demonstrated the usefulness of astronomical techniques to surveying and to mapmaking.

King was well aware of the limitations of astronomy in surveying. In

4 R.M. Stewart, 'Otto Klotz'
5 Report of C.H. McLeod, McGill University Observatory, Sessional Papers, Department of Marine and Fisheries, 1889, Appendix H
6 Dennis, 'A short history of the surveys,' 23

1906, as chief astronomer, he explained his reservations in his official report to the deputy minister of the Department of the Interior:

These astronomical determinations serve a useful purpose in the correction of maps, when the scale of these is not too large. For the control and checking of topographical surveys they are deficient. This is due to the fact that the astronomical and geographical coordinates of the same place are not necessarily, nor usually the same. The application of astronomy to topographical purposes proceeds on the assumption that the earth is a true spheroid and that the vertical line at a place (the direction of which it is the part of latitude and longitude observations to determine) is a true normal to it. The assumption is only approximately true: the irregularities of the earth, both above and beneath the surface, by their attractions, cause local deviations of the plumb line, so that the astronomical positions, though accurate in themselves within a few feet, may show a discrepancy in comparison with survey measurements of very considerable amounts.

There is still a wide field for astronomical determinations in Canada, both in correcting general maps compiled from local surveys not coordinated, and in affording new points of departure for geographical surveys in unsurveyed regions. They cannot serve as a control for topographical surveys of any degree of minuteness of detail. This is the function of the trigonometrical survey.[7]

Klotz, too, was aware of this problem, and as early as 1885, in his presidential address to the Association of Dominion Land Surveyors, he had discussed the advantages of a triangulation survey all over Canada.

But the success of the work along the BC railway belt established the importance of astronomy for the time being as the handmaid of the surveyor. It also stressed the need for a fixed observatory where time could be maintained and compared with the time as measured at a field location anywhere in Canada, thus determining the difference in longitude. Klotz and Drummond established Kamloops (based on Seattle) as a temporary base station during 1885 and 1886. Prior to this, Winnipeg (based on Chicago) had been used, and it was to be used again for the field seasons of 1887 and 1888. Early in 1887 Deville, King, and Klotz presented a joint submission to the deputy minister of the interior, A.M. Burgess, requesting that an observatory be established at Ottawa. Why Ottawa? It was the seat of government and the control centre for all

7 Report of the chief astronomer, Department of the Interior, 1906, Public Archives of Canada Sessional Papers 25A, 1907

federal operations. There were positive advantages to having the shortest possible lines of communication within the surveys Branch, and short lines of communication between all senior officials in the department and the office of the deputy minister. Burgess agreed with the submission, and so did the minister, the Hon. Thomas White. King and Klotz were permitted to visit Boston, Cambridge, New York, and Washington with a view to acquiring a large transit. They also visited observatories at Ann Arbor, Michigan, Madison, Wisconsin, and Northfield, Minnesota, to study their equipment.

Charles Carpmael, director of the Meteorological Observatory at Toronto, was quick to take action when he heard of plans for an Ottawa observatory. In a letter to the Hon. Geo. E. Foster, minister of marine and fisheries, he stated:

If this step is to be taken simply with the idea of having a station from which to determine longitudes, I would call to your attention the fact that there are already two observatories in the Dominion whose longitudes have been telegraphically determined, and which are thoroughly equipt with instruments necessary at a base station, and also possess competent observers. These are the McGill College Observatory, Montreal, and the Observatory at Toronto. Both the Observatories are under your department, that at Montreal partly, and that at Toronto completely under your control. I, and I am sure also Prof. McLeod of Montreal, would willingly assist in the determinations of such longitudes as may be required, so that there is no need to wait until a base station has been fixed at Ottawa before making accurate telegraphic determinations. Of course if there is any other reason for establishing an observatory at Ottawa I have nothing to say, but if it is solely for the purpose named, it seems to me that the existing observatories might be utilized.[8]

Carpmael was anxious that federal support of astronomy should not become diluted. His proposal, however, was not acted upon, because Deville, King, and Klotz were unanimous in their views about a base station in Ottawa. King had set up a temporary observing hut in his garden in 1887 to test the Russell alt-azimuth which he was to use in the field. Klotz also built one in his garden in Preston to gain experience with a zenith telescope.

Meanwhile the base station, though temporary in plan, continued to be Winnipeg. New and modern equipment, ordered as a result of the visit of

---

8 Carpmael to Foster, minister of marine and fisheries, 13 April 1887. Correspondence file, Meteorological Branch, Public Archives of Canada

King and Klotz to the United States, arrived in Winnipeg early in 1888. A transit of 3-inch aperture, 34-inch focal length, made by Thomas Cooke and Sons, was mounted in the new observing hut on Princess Street. A Brashear 8½-inch reflector with equatorial mount was made ready for occultation observations. A sidereal clock with electrical attachments was carefully mounted in the basement of the Clarendon Hotel. The acquisition of this equipment added weight to the argument for a permanent observatory in Ottawa.

That request was realized with the construction in 1890 of a modest frame observing hut on Cliff Street, at the edge of the escarpment overlooking the Ottawa River [Plates 21 and 22]. Klotz's diary for 24 June 1890, states: 'Next I saw Mr King, and with him inspected our new observatory here, a slight improvement over any temporary ones.' The transit instrument in this location became the reference meridian for the survey of Canada, being superseded only in 1905 with the inauguration of the Dominion Observatory on the Experimental Farm.

Field work on the precise measurement of base lines continued in Saskatchewan. The meridian from the head of Lake Timiskaming to James Bay, which forms part of the Ontario-Quebec interprovincial boundary, was connected by survey to Mattawa, and the longitude of Mattawa was connected by telegraph with Ottawa. In 1892 the department co-operated with the astronomer royal in the transatlantic longitude determination connecting Montreal with Greenwich.

That year King was named boundary commissioner to collaborate with his US counterpart, T.C. Mendenhall, later succeeded by W.W. Duffield, in the survey of the Canada-Alaska boundary. The three field seasons 1893 to 1895 were completely taken up with this problem, Klotz being in charge of one of the survey parties. In 1896, astronomical activity was resumed to the extent that Ottawa was connected with Montreal by King at Cliff Street and C.H. McLeod at McGill. King and Klotz then extended the connection to Winnipeg, thus linking the western survey to Greenwich. Port Stanley on Lake Erie was also connected with Ottawa to serve as a reference station for international boundary purposes.

The success with which the astronomical method had been applied to Canadian surveying prompted King to suggest that Canadians extend the longitude measurements across the Pacific and thereby complete the final circuit of the globe. 'In the summer of 1902,' King later reported, 'as the new Pacific cable was approaching completion, I pointed out to the Minister that an opportunity would thereby be afforded to extend our chain of longitudes from Vancouver (which had been connected with

Ottawa and Montreal in 1896) across the Pacific to Australia and New Zealand, and that the value to shipping of accurate positions on the Pacific would be even greater than on the Atlantic, while the completion of the first longitude circuit of the globe would be itself a notable achievement, which would be to the credit of Canada.'[9] The work was authorized and Klotz was placed in charge. During the spring and summer of 1903 he and his assistant, F.W.O. Werry, occupied alternate stations in the link: Vancouver to Fanning Island, Suva, Fiji, to Norfolk Island, to Queensland (Southport, Moreton Bay). Southport served as a temporary base station from which time was compared by telegraph with stations at Brisbane and Sydney. One detail of the work provided a pleasant surprise. 'Summer prevailed throughout the campaign' reported Klotz. 'Contrary to expectation, there was less annoyance from beetles and insects while observing than is found in Canada. A light in the open at night with us attracts multitudes of moths and beetles, and one's patience is sometimes sorely tried to have an unexpected visitor stake out a homestead on the nose just at the critical moment of the transit of a star.' The first satisfactory exchange between Southport and Sydney occurred 27 September 1903. Thus 'for the first time,' said Klotz in his report, 'longitude from the west clasped hands with longitude from the east, and the first astronomic girdle of the world was completed.' The achievement was even more noteworthy from the fact that 'the chain of successive differences of longitude carried from Greenwich through Canada was connected at Sydney with that carried eastward from Greenwich through Europe, India, etc., with a closing discrepancy in the circuit of the globe of only one-fifteenth of a second of time [about 84 feet]. As there are some twenty links in the whole chain, and the determination of the difference of longitude of each link consists of the comparison of time between its extremities, one can get some idea of the extreme accuracy of modern observations.'[10]

The observatory originally conceived by Deville, King, and Klotz was to be a modest site which might serve as a base station whose longitude would be determined with certainty. From here telegraphic longitudes could be extended to all parts of the Dominion served by the telegraph. In due course a large transit would be acquired and applied to the study of star

9 Report of the chief astronomer, Sessional Papers no. 25, Department of the Interior, 1905
10 A full report of the transpacific longitude determinations by Otto Klotz is given as an appendix to the report of the chief astronomer, 1906, Sessional Papers no. 25

positions, thus contributing to the international requirement for more precise information. The Cliff Street observatory answered this basic need. The equipment it contained is revealed in an inventory taken under the following unusual circumstances recorded by Klotz: 'What equipment we had in our old wooden observatory on Cliff Street is found from the following list made on the removal of the instruments on the night of the big fire in Hull and Ottawa, April 26, 1900: 2 switchboards, 1 watch chronometer, 1 old chronometer No. 81, 1 new Dent chronometer, striding level in case, old and new Cooke transits, galvanometer, relays, chronograph, Siemens switchboard, and instrument lamps.'[11] (There was no mention of fire damage to the observatory.)

When astronomical work resumed in 1896 after the Alaska boundary survey, it became apparent that the Cliff Street accommodation was but a stop gap. The observatory was a small frame shed, the primary clocks were in the old Supreme Court building, and the offices were in the Thistle Building on Wellington Street. The operations were scattered over three-quarters of a mile. A proper observatory was required where instruments could be mounted, tested, and put into service, where astronomical observations could be made, and where the chief astronomer and his staff would have adequate office space. It would then become possible to extend the range of operation from the utilitarian applications of surveying and timekeeping into broad avenues of astronomical research. Canada would take her place among the more advanced nations. It was time for a change.

Fortunately for W.F. King there was a sympathetic ear. The Honourable Clifford Sifton had become minister of the interior with the change of government in 1896. Two years later King submitted to him a proposal for a modest observatory, including with the memorandum the details collected by Klotz of over three hundred observatories in other parts of the world. Sifton asked what other observatories there were in Canada, and was told of the one in Montreal and the one in Toronto. Their work, said King, would not interfere with that contemplated in Ottawa.

From here the plans grew. The original proposal for a 10-inch equatorial telescope evolved into a 15-inch, the mechanical parts of which were built by Warner and Swasey of Cleveland, while the optics, grating, and so on were made by J.A. Brashear of Allegheny. The small building, which would have fitted on the knoll back of the Parliament Buildings overlook-

11 Klotz gives the steps leading up to the building and inauguration of the new observatory building in *J. Roy. Astron. Soc. Can.*, 13, no. 1 (1919)

ing the Ottawa River, became instead the spacious observatory on the high ground of the Experimental Farm [Plate 24]. Otto Klotz submitted a plan in which the front of the building formed a straight line; King's plan had a right angle; as a compromise the architect adopted an angle of 150 degrees between the two wings, and that was the way it was built. The first sod was turned in mid-summer 1902.

The observatory included the transit house, north and south azimuth mark buildings, a small observatory for stellar photometric work, a magnetic hut, the director's house, and a machine shop. The cost was as follows: observatory and auxiliary buildings $180,000, instrumental equipment $100,000, library, 9000 volumes, $30,000, making a total investment of $310,000. While the building was under construction and this sum, large in those days, was being spent, King prepared the following summary of the purposes of the observatory:

1 Primarily, the equipment of the Dominion Observatory is designed for the carrying out of definite lines of observation and investigation of a scientific character. These observations will not be astronomical only, but meteorological, seismological, spectroscopic, etc.

Such observations are for the general advancement of science, and, if conducted on a systematic plan, are of the highest value, though this value computed in dollars and cents may be remote. The absence of direct return from astronomical investigations calls for the aid of Government in a greater degree than do most other branches of science. State aid is given to astronomy by all civilized nations, and it is fitting that Canada should take part in the development of this branch of science.

2 The advantage to a nation in supporting such a science as astronomy is not confined to the ultimate results. In the very course of the work the men who do it derive special training, not only in the observing itself, but in the practical application of many branches of science which are closely bound up with modern astronomy.

This tends to develop a class of men of special training and knowledge who will be useful to the country, wherever accurate observation and investigation is required. For instance, explorations, geodetic work, etc., call for men of this stamp whose trained intelligence enables them to adapt their methods to overcoming the peculiar difficulties which they encounter.

3 A branch of the work of the Observatory is the determination of longitudes. The determinations which have already been made under this office have proved useful in the construction of maps, the correction of surveys, etc. They have from time to time been asked for by various departments of the public service. Under

the organization which the completion of the observatory will render possible, they can, it is believed, be made more rapidly and economically.

4 Another branch of the work will be the transmission of accurate time to the city and the public buildings.

5 Arrangements will be made for testing chronometers, a service frequently required by many departments, but for which there are now few facilities. Minor instruments can also be tested, such as sextants, aneroid barometers, etc.

6 An indirect advantage will be the public interest which it is hoped to arouse in astronomy and science generally by the exhibition of the instruments to visitors at stated times. In many countries, above all the United States, large contributions to scientific objects have been made by private persons. There seems to be no reason why similar benefactions should not be made in Canada, if public interest were directed towards these objects.[12]

In April 1905 the Dominion Observatory was inaugurated and went into operation. W.F. King became its director, while remaining chief astronomer of the Department of the Interior. Deville, who had been instrumental in forming the astronomical section within the Surveys Branch, culminating in the Cliff Street Observatory, now witnessed the beginning of an independent and thriving branch. It was the parting of the ways. He saw the observatory-controlled clocks installed in many of the government buildings, but never was one permitted to grace the walls of his office. He had a new transit installed at Cliff Street, and the pendulum clock which was used for testing surveyor's timepieces and which was verified by astronomic observations often more than a week apart, provided him with his own independent source of time.

The Cliff Street Observatory continued in use as a test site for another three decades after the new observatory was built on the Experimental Farm. The old wooden hut was replaced with a new structure about 1912; a cement flat on the west side was used for setting up surveying instruments and checking their performance; and a north azimuth mark was painted on the Alexandra interprovincial bridge near the Hull end. The Cliff Street site finally yielded to the Supreme Court building just before the outbreak of the second world war.

In 1902, before the construction of the new observatory, Robert Meldrum Stewart (1878–1954), winner of the gold medal in mathematics and physics at the University of Toronto, had joined the staff [Plate 25]. He had helped develop the embryo time service described in King's report.

12 King's summary of the purposes of the new observatory are recorded by Klotz, ibid.

When the move was made to the new building his appointment became permanent, and he was made superintendent of the Time Service. In 1924, following the death of Otto Klotz, he became the third director of the Dominion Observatory, and served in this office until his retirement in 1946. Stewart plays an important part in our story. As superintendent, he accomplished the difficult task of installing and debugging the large transit telescope called the meridian circle, and establishing an observational program to improve catalogue positions of those stars particularly useful to Canadian surveyors. He saw the introduction of radio time signals which freed the surveyor from the constraints imposed by the telegraph line, and which permitted the direct intercomparison of the time as determined by star observations at various observatories around the world. During his two decades as director, Stewart saw the pendulum timekeeper at its best and contributed his talents to the design of a time machine which mechanically produced correct mean solar time from the master sidereal pendulum. The rapid development of the quartz crystal oscillator had commenced to supplant the pendulum by the end of the second world war, and only six years after his retirement the pendulum clock at the observatory was also retired. Many of the events in the account that follows took place under Stewart's leadership.[13]

In summarizing the purposes of the Dominion Observatory in 1904, W.F. King had mentioned a local responsibility: transmitting accurate time to the city and its government buildings. This function remained a small but significant part of the observatory's operations until the 1950s and still continues, albeit in modified form. A characteristically experimental system of local time distribution had developed in Cliff Street days; it was later described by King in his report:

In the basement of the Supreme Court Building are set up two clocks, one regulated to sidereal and one to mean time.

The sidereal clock is connected by wire with a chronograph placed in the transit shed on Cliff Street, alongside the transit instrument. This enables star transits to be recorded upon the chronograph, and thereby the error of the sidereal clock to be determined.

The mean time and sidereal clocks are both connected by wire with the Langevin Block and this building.

Through the wires an impulse is given at each beat of the mean time pendulum

13 Miriam S. Burland, 'Robert Meldrum Stewart,' *J. Roy. Astron. Soc. Can.*, 49, no. 2 (1955) 64

to the pendulums of two clocks, one in each of the buildings mentioned. These impulses force the pendulums to beat synchronously with the master clock. The system by which this is effected is the invention of the late Professor A. Cornu, and the mechanism in use, including the master and the two controlled clocks, was made by Borrel, of Paris.[14] Each controlled clock carries a bar magnet at the lower end of its pendulum. This magnet passes, when the pendulum moves to the right through a solenoid which carries the controlling current ... When the pendulum swings to the left, the magnet passes through a copper cylinder, which has the purpose of damping the oscillations, and thereby rendering the controlling impulse paramount over the natural period of the pendulum.

The two controlled clocks each drive, by independent circuits, a number of dials operated each minute by an electromagnet ...

On the switchboard there is ... a switch by means of which a current can be sent in either direction to regulate the mean time master clock in the basement of the Supreme Court building. This clock is provided with a permanent magnet on the pendulum, which swings with it over an electromagnet placed below. The passage of a current in a direction to make the electromagnet attract the permanent magnet accelerates the pendulum; if in the opposite direction, retards it ...

This apparatus enables all the operations to be conducted in the office, except the observations themselves, and thereby obviates as much as possible the inconvenience of having observing station, clock room and office so far apart. The observer takes his observations at night, brings his chronograph record sheet to the office in the morning, and there works out his observations. This gives the error of the sidereal clock. A comparison is then made by means of the switchboard between the sidereal and the mean time clock. If the mean time clock is found to be in error, the small correction necessary is made by turning the switch to the right or left.

This service was put in experimentally. It has proved generally satisfactory, though it will receive certain modifications when the completion of the new building renders possible a more comprehensive scheme.[15]

14 Cornu method: Cornu, a clockmaker, described a method for holding pendulums in synchronism with a master clock. Each controlled clock carried a bar magnet at the lower end of its pendulum. At one end of the swing the bar passed through a solenoid which carried the controlling pulse, and at the other end of the swing it passed through a copper cylinder which dampened the oscillations. The controlling impulse kept the pendulum in operation, regardless of its own natural period, and without the impulse the controlled pendulum came to a stop because of the copper cylinder. The Cornu method was discontinued in favour of one which permitted the controlled pendulum to continue operating freely with the control removed.
15 Report of the chief astronomer, Department of the Interior, 1905, Public Archives of Canada

The tradition of service to the community was inherited by the Dominion Observatory at its inauguration in 1905. Most of the work at the new institution was to be of a research nature, with little if any immediate return to the people of Canada. Time, however, is a commodity required in every well-regulated society, from the city of Ottawa to the wider community of Canada. The telegraph served that portion of Canada adjacent to the railway, and later, when radio permeated the land, time was synchronized via radio and thereby television. At first it was radio from outside Canada, broadcasting time signals from other national time laboratories, that served as a link between Ottawa and the field stations. Then Canada's own short wave station, CHU, was developed as a service to as wide an audience, both national and international, as could profit by it. But we are getting ahead of our story.

The time room on the main floor of the new observatory contained the secondary pendulums, both mean time and sidereal, which were required for the operation of the time service [Plate 43]. The various circuits radiated from a marble control panel, at the base of which were two glass-faced shelves containing the necessary relays. Above the shelves was a row of eighteen jacks into which could be plugged the cord from a test meter mounted centrally on the panel. A variety of switches made it possible to change the control from one pendulum to another.

In 1905, the downtown time service included four government buildings, the East Block, the West Block, the House of Commons, and the Langevin Block. Secondary master clocks in each building controlled a number of dials the power for the output of each clock being supplied by a 24-volt bank of Edison cells, with motor generator at each site to charge the cells each week. Other services supplied from the time room included seconds pulses to the chronograph for transit work; pulses to the seismograph shutters; signals by the telegraph line to the local offices of the Great Northwestern (later CN) and Canadian Pacific telegraph rooms; and minute pulses for the tower clock and mean time dials in the observatory. Gradually the time service was extended to other government buildings, from six secondary master clocks controlling 214 minute dials in 1905 to sixteen secondaries controlling 682 minute dials in 1930. The number of government buildings for which a time service was provided had increased from four in 1905 to twelve by 1939.

A program clock, in the form of a punched tape controlled by one of the secondary mean time pendulums of the time room, gave the hourly pulse to the seismic recorders, the daily signal to the telegraph companies, and other services of a repetitive nature. The observatory directly controlled the tower clocks on the Post Office and the Peace Tower. Also, by 1930 clock beats could be switched onto the telephone line for the benefit of jewellers, surveyors, and others who might call in for the correct time.

The number of dials in Ottawa government buildings controlled and maintained by the Dominion Observatory grew to a maximum of about 750 during the second world war, not including one hundred or more dials in the Centre Block on Parliament Hill whose maintenance had been transferred to the resident engineer. After the war some of the installations had aged almost beyond repair, and they were gradually replaced by electric wall clocks during renovations. Newer buildings were equipped either with electric wall clocks operating directly off the alternating current supply or with co-ordinated dials controlled by a master clock in the building. By the end of the 1950s the observatory had divested itself of the maintenance of all clocks and dials in the public buildings. An hourly synchronization pulse continued in service, primarily for the Parliament Buildings, and after the Time Service was amalgamated with NRC in 1971 this was replaced by a continuous seconds pulse with a voice announcement each minute. This was a more useful signal to check the clock in the Peace Tower and to provide the time marks in the recording of *Hansard*.

Let us pause to recall that the observatory was established to serve as the base station for the survey of Canada, where precise timekeepers, checked regularly by transit observations of stars, were stored. Every season the surveyors and their parties, each equipped with a transit and sophisticated portable chronometers, would occupy their assigned stations. At each station the local astronomical time would be determined and compared with observatory time, either by telegraph or radio, and thereby the difference in longitude would be computed. At the end of the season each party would return all gear for maintenance, repair, and storage. The chronometers would go to the fully modern watch repair shop [Plate 46] where facilities were available to clean and regulate both sidereal and mean solar timepieces. Because it was a specialty shop, program clocks from other federal services, such as those used in navigation buoys, were brought in to be repaired and regulated too.

After the second world war, the Geodetic Survey, which had been inaugurated as a part of the observatory, became part of the Surveys and Mapping Branch of the Department of Energy Mines and Resources, and the watch repair shop moved with it.

4

The evolution of
astronomical techniques

When the Dominion Observatory was inaugurated on the corner of the Experimental Farm southwest of Ottawa in 1905, the Canadian reference meridian was moved there from Cliff Street. The transit annex of the observatory was not completed until a couple of years later; during the interval a temporary transit hut was occupied to the east of the observatory building, and its observing pier defined the reference meridian. Then, when the transit annex was completed and the meridian circle telescope placed in operation, the reference meridian was relocated as the north-south line through its optical axis, originally determined as 5^h 02^m $51^s.983$, subsequently modified to 5^h 02^m $51^s.940$.

A transit instrument, originally designed by Ole Roemer of Denmark in 1690, is a relatively small telescope mounted on a horizontal east-west axis so that it has freedom of motion in one plane only, the vertical plane of the meridian. It can normally be pointed in any direction along the meridian, from north to south, and then depends upon earth rotation for motion from west to east. A star on which the telescope has been set will be seen to drift from east to west across the field of view. At the equator a star will appear to move rapidly. If one sets on a star away from the equator, it will appear to move more slowly. At the pole a star will appear to be stationary. The earth will have made one complete rotation on its axis when, exactly twenty-four hours later, the same star is seen to pass again through the centre of the line of sight. Due account of course must be made for precession and proper motion, precession being the steady drift eastward of the stars because of a westward motion of the equinoxes and proper motion the change from the star's own movement in space.

A great deal of preliminary work must be done before the stars can be used to determine the time. Hours of painstaking observing are required to 'map' the sky, to determine the positions of a representative group of stars within the slowly changing co-ordinate system described by the equator and the vernal equinox. A star catalogue in use today is thus the result of the accumulated effort of many dedicated individuals who have spent the quiet watches of the night following the majestic parade star by star, measuring the elevation and marking down the time of transit of each one.

The transit instruments used in the early survey of the Canadian west were equipped with vertical threads, or similar markings on a glass reticule, and the star would be observed to pass each thread during its transit. Illumination of the threads was provided by a small amount of light introduced through one end of the hollow horizontal axis and reflected towards the eyepiece by a prism or small mirror. Otherwise the threads would not be seen. For a bright star the light could be bright, but for fainter stars the light had to be dimmed, up to the point where the threads or lines were just visible. As a result, transit observations of stars by eye were limited to the brighter stars.

A clock or chronometer or watch is an essential tool for the astronomer, so that he can determine the time of transit of each star, that is, the instant that the star is on the optical axis of the telescope. Conversely, timing the actual transit of stars whose positions have been well established enables him to measure the error of the clock, and thereby to establish correct local time.

The first improvement in meridian observing came with the invention in 1850 by W.C. Bond at the Harvard College Observatory of the spring governor, a gravity-driven drum chronograph on which the beats of the clock, together with the electrical impulses transmitted by the key in the hand of the observer, were recorded [Plate 26]. With this convenient recording device, the observer watched the star approach each thread and, at the mid-point of transit of the thread, pressed the key. Prior to this he wrote down the time. Bond described the advantages of his device. 'In three most important requisites it has unquestionably the advantage over any of the plans hitherto used; it is more accurate in its results; it is superior in the point of convenience, and in this respect recommends itself to the observer, relieving him from much labor, and contributing to the ease and comparative comfort with which the work can be prosecuted; lastly, the time necessary for completing an observation is greatly shortened.'[1]

Learning to use a transit telescope has been described by Otto Klotz,

1 Annals of the Astronomical Observatory of Harvard College, Vol. 1, pt 11, p. 5. Harvard College Archives.

who had his introduction into its mysteries when he was named astronomer in charge of the longitude and latitude work along the Railway Belt in British Columbia. Seattle was selected as the starting point because the Canadian survey had not extended far enough west. The timepiece was a box chronometer beating half seconds, and the transit had five threads. Transit observations were made by the 'eye and ear' method. The astronomer would count to himself the seconds and half seconds as he heard them, holding the watch to his ear; at the same time he would watch the star as it passed behind each thread in succession, recording the times. Although he was skilled in the use of the theodolite and other tools of a surveyor, this was his first experience with a transit telescope. 'How well I recollect' wrote Klotz 'how the perspiration ran down my face and neck while observing transits at Seattle, my first station [June 1885]. Picking up the beat of the chronometer; carrying it mentally foward while watching the transit; hurriedly jotting down the beat, still carrying on the counting, for to look at the chronometer would probably mean the loss of catching the transit of the next thread, was strenuous work ... With continued observations some composure was attained, and the transits were found to take place not only at the beats and midway, but at other divisions.' Later that same season he could report: 'Composed I sat at the eyepiece. The chronometer beats were slowly (apparently) running their course and the mind could easily, yet unconsciously subdivide the time intervals ... the tenth of an interval or twentieth of a second ... occasionally being estimated.'[2] Klotz continued: 'To test the accuracy of the longitudes determined by the eye and ear method, one of the stations was many years afterwards occupied by another observer supplied with the most modern outfit; Cooke transit with transit micrometer; chronometer with electric attachment and drum chronograph; observing seven stars instead of just five for one position of the clamp, besides using more zenith stars and fewer equatorials than formerly. The longitude obtained agreed within a few hundredths of a second with that of the eye and ear method.'[3] This was perhaps a reference to the remarkable work of William Ogilvie on the Alaska boundary during the winter of 1887–8, followed by that of F.A. McDiarmid in 1906, in which the difference as marked on the ground was only 218 feet.

The next big advance in observing technique was the travelling wire micrometer, designed by J.G. Repsold towards the end of the nineteenth

2 Klotz, *J. Roy. Astron. Soc. Can.*, 13, no. 5 (1919) 288
3 Don Thomson, *Men and Meridians*, Vol. 2, 196

century. Described also as 'the impersonal micrometer,' it was hailed as the device which would free the observer of his 'personal equation': he would no longer be guilty of consistently pressing his observing key a little early or a little late compared to another observer. Always at the beginning and end of a field season the field man was required to observe along with the base station astronomer, each using his own transit, to measure this systematic difference. Now that was no longer required. A single wire, or a closely spaced pair of wires, was fastened to a slide which could be made to move in an east-west direction by a screw terminated at each end by a small hand wheel. The eyepiece was carried on the same slide so that the movable wires appeared always in the centre of the field of view. Now the astronomer watched as each star moved into the field of view. Then when the star was properly bisected by the wire, or was at the bisection position between the pair of wires, he would turn the hand wheels so as to hold the star at that attitude till it had moved across the field of transit. The hand wheels also drove a commutator wheel against which a brush made contact with each of the strips embedded in the wheel. These contacts were recorded along with the clock beats on the drum chronograph to mark the time that the star was in various positions during transit. They effectively gave the time, according to the local observing clock, when the star was on the meridian, that is, on the optical axis of the instruments.

R.M. Stewart at one time believed that he had a small and fairly constant personal equation with the old fixed-wire method. In a series of tests to measure the personal equation between himself and F.A. McDiarmid, he used an instrument equipped with a glass reticule while McDiarmid used one equipped with a transit micrometer. The difference between the two averaged out to about 0.35 seconds. Then Stewart ran another series of observations in which he himself alternated in the use of the two transit instruments. To his surprise several nights work showed the same difference of 0.35 seconds in his clock corrections, convincing him completely that he had a large personal equation with the fixed reticule, and that the transit micrometer eliminated from the results most of the effects of the observer's personal equation.[4]

Yet there was still the tendency for the observer to lead or lag systematically as he followed the centre of the star across the field. This was overcome in the field type of transit (the Cooke or Heyde) by reversing it on its pivots at the mid-point of the transit. Then the star would be seen to

4 F.A. McDiarmid, *J. Roy. Astron. Soc. Can.*, 2, no. 2 (1908) 93

enter and leave through the same optical side of the instrument, and also the apparent direction of motion of the star would be reversed. The first effect, i.e. using the same slide of the little telescope to observe the entrance and departure of the star, cancelled the effect of any collimation error. The second effect, namely reversing the direction, meant that the contacts from the commutator wheel were the same in reverse order, except for the width of the contact strips. The first and last contact were therefore equally distant from the time of mid-transit, and the second and second last, and so on. The contacts were in groups of twenty, two such groups occurring on each side of centre. It was usual practice to use the outer two groups of twenty, except for slow-moving polars, leaving a gap during which the star traversed the centre of the field. There was ample time during this gap to reverse the telescope on its pivots and reset it on the star.

The larger meridian circle transit telescope [Plate 28], whose reversal on its pivots required upwards of half an hour and in fact was routinely done only once or twice a year, was equipped with a dove prism in the optical path of the eyepiece. Turning the prism through 90 degrees reversed the optics by 180 degrees, and gave the impression of the star reversing its direction of motion, though of course the micrometer hand wheels had to continue to turn in the same direction. Thus the prism was turned as the star passed through the centre of the field. When the meridian circle was reversed, the clamp which held the telescope firmly once it was set on a star also changed sides. So the setting of the telescope at any time was either 'clamp east' or 'clamp west.' It usually remained in one clamp for a year, being reversed for the following year so that the stars of each program would in due course be observed in both positions of the telescope.

Great care was taken to measure and account for the personal equation, especially during the world longitude campaign of 1933. A 'personal equation machine' was built to simulate a star moving at three different speeds which could be selected by the observer, one speed to simulate a fast-moving star of zero degrees declination, one for a mid-declination of 45 degrees, and one for a slower-moving polar at 72 degrees. Contacts were made by the machine as the star (small light bulb behind a pin hole) was driven from right to left and then back again. These were compared to the contacts recorded by the eyepiece micrometer.

Some doubt seems to exist concerning the validity of the personal equation machine. C.C. Smith noted that in the 1933 world longitude observations in Vancouver he and H.S. Swinburn, both from the Dominion Observatory, had a relative personal equation of −0.10 second as revealed consistently by the clock corrections derived from their separate observations. The personal equation machine gave a value of only −0.03 second. It was hardly to be expected that the machine could simulate actual conditions of observing. The Ottawa machine was placed in the south azimuth hut 150 feet away and viewed through a collimating lens. All the viewing therefore was done with the telescope in a horizontal position. One might have thought that the Vancouver measurements would have been more favourable since both observers used broken-type transit instruments with a fixed viewing position through one end of the axis. However, the two transits differed in both aperture and focal length, the one being a Cooke of 3 inches aperture and 3 feet focal length [Plate 32], the other a Heyde of 2 inches aperture and about 21 inches focal length, while the machine was mounted for use only with the Cooke. In the years after the second world war when several observers made use of the Cooke for time determinations at the observatory no attempt was made to measure and apply a personal equation. More weight was given to the internal agreement revealed by the individual stars, and to the fit of the night's work when compared with the clock corrections obtained on previous and subsequent nights.

When King and Klotz had formulated plans for a Dominion Observatory in which provision would be made for research, a logical research area was that of positional or fundamental astronomy. Of Canada's very long international boundary with the United States, half is an east-west line defined by the 49th parallel of latitude. Its position had been delineated astronomically, but using stars whose positions were not too well defined. Thus it was desirable to incorporate in the new building a larger transit equipped with accurate measuring facilities so that an accurate catalogue of these latitude stars could be made. It was recognized, of course, that improved star positions were also of practical value in the measurement of longitude, azimuth, and time. Beyond the local advantages of a good time service, a program of fundamental astronomy would provide reference stars for research in astrophysics and stellar photography, yield evidence concerning variations of latitude, longitude, and time, and of course provide the information from which to derive proper motions of the individual stars.

The meridian circle, purchased from Troughton and Simms [Plate 28], arrived in October 1907, but not until January 1911 was it finally placed in operation. A description by R.M. Stewart in his annual report of 1908 is as follows:

The telescope is of six inches aperture and about seven feet focal length; the field contains six vertical threads and two horizontal ones, in addition to the moveable micrometer threads; the right ascension micrometer is supplied with the Repsold automatic registering device. The field illumination is provided for by an annular reflector in the axis; bright wire illumination is effected by four small electric lamps inside the tube near the eye end. [This was later changed to dark wire illumination with a small lamp near the objective.] There are two circles [36 inches in diameter], each graduated to every five minutes, one being fixed in position on the axis, the other moveable. They are read by four microscopes each, two pointer-microscopes (one for each circle) being provided for reading to the nearest five-minute division. There is an end-thrust bearing at each end of the axis, one being fixed, while the other is tightened by two nuts pressing against coil springs; this ensures the constancy of the position of the telescope with respect to the standards, so that the division marks may always be in focus in the reading microscopes. There are two collimating telescopes each of four inches aperture and about four and a half feet focal length. For reversing the telescope a reversing carriage is provided [Plate 27], which runs on rails between the piers. The level is read by nadir observations on a circular mercury trough with the usual [Bohen-berger] collimating eyepiece.[5]

A platform mounted on top of the reversing carriage served as a stand from which the observer could look down through the telescope to see the micrometer wires reflected back from the mercury basin.

It was indeed a splendid instrument. The delay of more than three years required to make it ready for operation was a result of flaws in both manufacture and installation. The large graduated circles were returned to the maker because they were warped slightly in transit. The circle-reading microscopes had to be remounted because they were unstable. The pivots turned out to be soft and had to be renewed, and the counter-poises had to be completely remodelled. Iron shutters, which had been installed originally to cover the slots of the transit room, warped so badly they no longer kept out the rain or snow, and had to be replaced with wooden shutters. Perhaps worst of all, the cement piers had to be completely rebuilt so that they penetrated more deeply below the frost line. Furthermore, they had to be protected with adequate drainage. All this was done under the watchful eye of R.M. Stewart.

Finally, with everything in readiness, the collimating piers and the azimuth mark piers completed and the meridian circle almost completely

5 R.M. Stewart in Report of the chief astronomer for 1908, Appendix 3.

rebuilt, the first program was prepared. In his unpublished report of 1930, C.C. ('Charlie') Smith, chief of the Time Service, wrote:

With the purpose of confining the observations to those stars chiefly which were being used, or likely to be used in latitude and longitude observations for geographical positions in northern latitudes, a list comprising 3162 stars was compiled suitable for latitude observations. This list comprised, with some exceptions, the following stars: –

a / All those north of 20° declination, and many between 10° and 20° in Boss' Preliminary General Catalogue for 1900.

b / All those between the same limits of declination in Ambronn's Sternverzeichnis for 1900.

c / Such additional stars as had at any time been used by this observatory in latitude observations.

The exceptions were stars fainter than magnitude 7.5, and close unequal doubles considered unsuitable for measurement. The list included 255 standard stars, 180 fundamental stars and 21 azimuth stars to be observed.

Each meridian circle observation required the following operations – setting the telescope ready for the appearance of the star, reading the circle microscopes, observing the star in right ascension and in zenith distance,[6] reading the temperature, and, in addition the necessary readings for collimation, level, nadir and azimuth. Each observation required in computation 33 operations in right ascension and 36 operations in declination, and the final discussion of the results, which can only be done after the whole program is completely observed, is involved and lengthy. A total of 28,000 observations were made of the stars on this list. Incidentally that has required over one and one-half million computations. Let anyone who lacks patience and demands immediate results beware of undertaking meridian observations on a list of stars. The observations were completed in 1923.

The values of the micrometer screws and the series of observations necessary for the determination of pivot errors, were made in 1910, 1912 and 1916, and the readings for the corrections to the declination circle graduations in 1920. This latter series of measurements required a total of 22,000 readings. In Greenwich and Paris Observatories, where similar observations have been carried on for over two centuries, the multitudinous observations necessary for defining star posi-

6 Observing the star in right ascension refers to the transit observation described previously. Observing the star in zenith distance means determining the distance from the overhead position, or the elevation of the star, by reading the setting circle attached to the telescope's axis as well as the vertical micrometer adjustment of the eyepiece which brings the star to the centre of the horizontal pair of spider lines. The time of transit and the elevation of a star describe its position in the sky relative to all the other stars observed.

tions have ceased to be terrifying. Actually, to one who enjoys measuring, the work is not distasteful and the results are of great interest and are necessary.

C.C. Smith was away from the observatory between 1912 and 1919, and W.S. McClenahan was on military duty for three years. The observing team was for a while reduced to three persons, R.M. Stewart, D.B. Nugent, and R.J. McDiarmid. To these five men goes the credit for taking observations and doing a great part of the computing. Others who assisted in the computation were W.C. Jacques, E.C. Arbogast, and H.S. Swinburn, while Dave Robertson assisted in reading the chronograph records. Continued Smith:

By the time the Latitude Program was completed the meridian circle and its accessory equipment were in excellent condition; the different constants of the instrument had been almost all well determined and the experience gained was of very considerable benefit.

At the meeting of the International Astronomical Union at Rome in 1922, it had been recommended by the Commission of Meridian Astronomy that such observatories as had the suitable equipment should undertake the fundamental observation of part of a list of 3064 fundamental stars published in the *Connaissance des Temps* for the year 1914. It was decided, therefore, that the Ottawa meridian circle could render this service to astronomy and that it was our duty to do our part as one of the national observatories of the world in taking on the observing of such of these stars as could be most profitably observed here. The stars on the Ottawa list are known as the Backlund-Hough stars [so called after the two astronomers whose work was used to compile the *Connaissance des Temps* list] and comprise a list of 1368 stars. In addition to these there are, of course, the 180 clock stars, 21 azimuth stars, the sun, moon and planets.

The intention was to observe these stars twice in each clamp and then to interchange the eyepiece and object glass and repeat, a total of eight observations. The first half of the program was completed in 1935, and the second half then commenced.

Several events combined to reduce the scope of this well-conceived plan. When C.C. Smith retired at the end of 1937 the program was deprived of one of its most enthusiastic observers. The same year saw the publication of Benjamin Boss's *General Catalogue* of the positions and proper motions of 33,342 stars. It seemed as if the job had been done, and general interest in positional astronomy tended to flag. The Ottawa program was reduced in scope by dropping observations of the sun, moon, and planets. The position of the ecliptic, the path described by the apparent sun in its annual journey around the earth, and hence that of the stellar co-ordinate system, could no longer be derived from the Ottawa observations. Instead the stellar positions were determined with reference to the framework of standard stars. By the outbreak of the second world war, R.J. McDiarmid and H.S. Swinburn alone remained to continue the meridian circle program, while W.S. McClenahan was responsible for time determinations using the broken-type Cooke transit.

Charlie Smith, who had joined the Observatory in 1908, preferred to work on the practical or observational problems of his profession [Plate 30].[7] During a seven-year absence he was successfully self-employed as a surveyor on the west coast. His love for music often found him at his piano in the small hours of the morning as he relaxed after a session of observing, and occasionally his rich bass voice could be heard coming through the open shutters of the transit room in song as he waited in the darkness for the next star to appear. He was succeeded as chief of the Time Service in 1938 by D.B. ('Bert') Nugent, a quietly friendly scientist who had joined the staff in 1907 and made a notable contribution as an observer during the pioneer days and throughout the first program of the meridian circle [Plate 31]. In 1923 he had been assigned to work up the clock corrections from the meridian circle transit observations and supervise the maintenance of the primary clocks of the observatory as well as the downtown time service. Ill health forced his retirement in 1944.

W.S. ('Bill') McClenahan, who succeeded D.B. Nugent, was one of those individuals who seemed to be good at everything he undertook [Plate 26]. Abreast of the developments in his chosen field of astronometry, he kept careful notes of what he read and studied. He enjoyed sports, and as a group leader he was very generous in giving encouragement; yet the aim of those who worked with him was always to try to do as well. McClenahan had the distinction of being able to observe a set of stars, reverse the meridian circle single-handedly, and then accumulate a second set. The reversal of the meridian circle on its pivots involved moving the hoist into position, carefully cranking the telescope up, moving the hoist with its load into the clear so that the telescope could be turned horizontally through 180°, then carefully returning the telescope to its pivots. It was a task usually done in daytime and by two people. Two star sets could be accumulated only on a long winter night, because the reading of the instrumental constants, added to the time required for transit observa-

7 D.B. Nugent, 'Charles Campbell Smith'

tions, meant that a single set took about $4^{1}/_{2}$ hours. Two sets meant a long night.

During its half century of operation, the meridian circle was applied to six observational programs. It had been the original intention to complete the discussion of each program and make the results available for international use soon after the conclusion of observing. Staff limitations, depression, and war all contributed to frustrate this intention. Not till war was over and new staff was available under the leadership of McClenahan were the original observations dating back to 1911 published, largely due to the assistance given by E.G. Woolsey and R.W. Tanner. Also, electronic methods of data reduction were beginning to permit computations to keep up with observations, so that final discussion and publication could be accomplished within two years.

Several improvements were incorporated in the meridian circle in the post-war years. The mercury reflector was floated in a basin of mercury to reduce vibrations when the nadir was being read. Thirty-five millimetre cameras of local design replaced the microscopes for circle registration, and E.G. Woolsey designed a projection machine for measuring the film. A reversible motor replaced the hand crank for the declination setting, though the fine vertical adjustment to bring the star within the horizontal wires of the eyepiece continued to be by means of a hand wheel on the eyepiece assembly. The mirror transit telescope [Plate 34], intended as a successor to the meridian circle, was to be completely impersonal, with photographic registration of stars and instrumental settings, remote servo control, and on-line punch card facilities for all the routine calculations. Unfortunately it never got beyond the experimental stage, and work was discontinued when it was realized that extensive funds would be required for a major redesign and relocation.

The meridian circle transit was used for the determination of astronomical time up to the beginning of 1935, after which reliance for timekeeping was placed on the broken Cooke. The meridian circle, because of its greater focal length, gave results much more consistent from star to star on any one night. On the other hand the Cooke yielded more consistent night-to-night clock corrections because of the reversal at mid-transit of the smaller telescope and the consequent cancelling of the collimation error, and also because of the reading of the striding level before and after the 180° reversal. This was well demonstrated by a comparison of the two instruments over a period of years.

C.C. Smith had recognized that the smaller Cooke transit was better than the meridian circle for night-to-night clock corrections. He was also aware that the small instrument with a straight tube has certain disadvantages that would be obviated by breaking the light beam by a mirror or prism at the centre of the little telescope, and directing it at right angles through the hollow axis to an eyepiece attached to an extension of one of the pivots. The following were the advantages of this 'broken'-type transit telescope: 1/ the observer's eye remained at the same convenient position regardless of the pointing of the telescope, and hence the errors due to the position of the observer north and south of the zenith were avoided; 2/ the level hung directly under the rotation axis and was read on the two settings for every star; 3/ the instrument was very stable and easy to operate; 4/ zenith stars could be observed, thereby reducing the azimuth correction. It was impossible to observe a zenith star with a straight instrument because there was no room for the head of the observer between the telescope and the pier. From twenty degrees north to twenty degrees south of the zenith, there was a blind zone on the straight type.

There happened to be an extra straight Cooke transit of the required rugged design, which, under Smith's instruction, was reconstructed in the observatory machine shop [Plates 23 and 32].[8] The result was an excellent broken-type transit at very little expense available in time for the instrument to be used in Vancouver by Smith in the 1933 world longitude campaign. In 1935 it came into regular use for the determination of the clock correction at Ottawa. Smith used it as long as he was with the Time Service, and on his retirement in December 1937 Bill McClenahan took over its operation.

For nearly a decade, including the war years, the Time Service at Ottawa depended on the clock corrections provided by McClenahan. On 28 August 1941, the Dominion Observatory time was designated by order-in-council as official time for Dominion official purposes. Although the Saint John observatory was not closed until 1951, the Dominion Observatory was the only government agency equipped to become the custodian of official time, and now, four and a half years after the 1936 reorganization of government departments, it was recognized as such.

The post-war years produced a quickening of activity at the observatory. R.M. Stewart, who had agreed to remain at his post two years beyond normal age for retirement, and who for patriotic reasons had declined even to take time off for annual leave, retired in 1946. Dr C.S. Beals, who had distinguished himself for his contribution to astronomy at the Dominion Astrophysical Observatory in Victoria, succeeded Stewart as Domin-

8 C.C. Smith, 'Converting a transit from a straight to a broken type'

ion astronomer [Plate 35]. In that capacity Beals continued to maintain an active interest in astrophysics, and indeed made some significant contributions while at the same time keeping up with an increasing load of administrative duties. Then Beals effectively commenced a new career and won international recognition for his investigations of fossil meteor craters in Canada. Perhaps this helped to focus attention within the branch on the geophysical areas of research at the Observatory, namely gravity, seismology, and geomagnetism. Certainly under his direction the geophysical effort expanded till it more than doubled in size the total astronomical manpower within the Observatories Branch. When government responsibility for astronomy was much later consolidated under the National Research Council in 1970, the geophysical remnant of the former Observatories Branch was of sufficient size and competence to form the Earth Physics Branch of the Department of Energy, Mines and Resources.

A similar surge of encouragement and enthusiasm was injected into the astronomical effort at Ottawa, solar physics, meteor physics, and positional astronomy all receiving encouragement to adopt modern methods, plan new and imaginative programs, and acquire the manpower to replace the shrinkage which had occurred during the 1930s and second world war.

I had joined the time service in 1930, having obtained a BA from the University of Manitoba the previous year. During the 1930s, under the guidance of J.P. Henderson, I had worked on the time signal program and helped in the gradual development of the small transmitters which developed into CHU. When I reported back to the Dominion Observatory in 1945 after three years in the RCAF, I was assigned by Bill McClenahan, who was chief of the division, to a regular routine of observing with the broken Cooke transit [Plate 33]. Occasionally, on cold winter nights, the smoke from the chimney of the Forage Division building of the Experimental Farm or from that of the Civic Hospital, both in the direction of the prevailing fair weather westerly air flow, would momentarily hide a star in transit. But one would maintain the steady motion of the micrometer wheels and feel gratified if on its reappearance the star was still at the bisection point between the moving wires. At times the seeing was decidedly bad, the star appearing to move erratically, sometimes jumping beyond the micrometer wires. If the weather dropped to minus 15°F (minus 26°C), it was considered too cold, both the moving parts of the transit and the observer's fingers being too stiff for proper operation of either the meridian circle or the Cooke. Before that critical temperature arrived, however, one's breathing frosted the eyepiece and the moving parts of the micrometer box. Wiping the eyepiece with alcohol cleared the frost. Then, to prevent more from forming during transit, it was common practice to hold one's breath while keeping one's eye fixed on the eyepiece, then to turn away with lungs fairly bursting to prepare for the second half of the transit. The cold always seemed to seep through to the marrow, in spite of frequent retreats to the warmth of the time room. I remember racing up and down the stairs from basement to dome to restore the circulation, and the brisk two-mile walk home was just about right.

The physical discomforts of cold weather observing largely disappeared with the acquisition of heated flying suits in 1948. I recall commencing an observing session with the heavy feeling of a cold; then as the heat of the flying suit kept me comfortable in spite of the freezing weather the heaviness disappeared, the cold fresh air cleared my head and lungs, and next day I felt fine. Of course the flying suits were a potential hazard. One evening R.W. Tanner rushed into the time room from the meridian circle exclaiming 'I'm on fire!' I hastily helped him out of the suit and found that the wiring had developed a hot spot and had burned a hole through his underwear to his thigh. Apparently it was the only incident of its kind.

On balmy summer evenings in Ottawa mosquitoes readily found their way through the open shutters; but summer observing generally involved little in the way of physical discomfort. The best seeing occurred when a light haze covered the sky, a condition that occasionally prevailed in the spring and summer. It was always a thrill to see a star move steadily across the field of the transit and keep it centrally between the micrometer wires. Then one looked forward to working up the time set next day, feeling sure that it must be a good one.

A 'time set' consisted of twelve stars, two of which were polars. The ten time stars were selected as close to the zenith as possible in order to reduce the azimuth correction to a minimum. Azimuth error, a result of the axis not being in a true east-west direction, was measured from the stars themselves, the polars combining with the southern stars to indicate how much the instrument might be pointing off from the meridian. At the mid-point of each transit the instrument was reversed on its pivots, and the level was read in both positions. A good time set, generally speaking, was one in which clock corrections from the individual stars agreed with each other to within one-tenth of a second. Sometimes, when conditions were poor, the spread could be as much as three-tenths of a second. Bill

McClenahan had one set with an internal agreement of seven hundredths of a second, a record.

The Cooke transit was a sturdy, well-built instrument, mounted on a heavy metal plate on a cement pier which extended down to a footing well below the frost line. It seemed to be unaffected by the weight of the observer's hands on the micrometer hand wheels, though it must be admitted that the operation was done with as light a touch as possible. On 21 August 1949, as J.P. Henderson was observing, he noticed that the bubble of the striding level moved full scale and was starting back. He put his weight against the heavy uprights to see if he could detect some instability and arrest the motion, but to no avail. He then carefully observed the extent of the excursion, the interval of the swing, as well as the time. Next day the Seismology Division confirmed that the earth motions indicated by the excursions of the striding level of the Cooke corresponded exactly with the seismic recordings of an earthquake that had occurred offshore from Prince Rupert, BC. It is not known whether an earthquake has ever been recorded before or since by a transit instrument!

After the second world war it became increasingly evident that the transit observations with the Cooke formed the weakest link in the determination and maintenance of fundamental time at Ottawa. Crystal clocks which supplemented the pendulum clock were exhibiting a day-to-day precision of a millisecond or two, whereas the results from a time set carried an uncertainty of about five one-hundredths of a second. Evidence from the US Naval Observatory at Washington was that the results from the photographic zenith tube (PZT) were ten times more precise than those from a transit instrument.

Before the outbreak of the war, the PZT had received some attention at Ottawa, and indeed preliminary steps had been taken to compile a list of zenith stars suitable for the latitude of Ottawa. Soon after C.S. Beals became director in 1946, McClenahan was commissioned to visit Washington and to discuss with experts there the properties of the PZT. Plans at that time were well advanced in the United States for the construction of a new instrument, to be located at Richmond, Florida. The original one in Washington had been built by F.E. Ross in 1908 to detect latitude variation, and had been modified twenty years later so that both time and latitude could be determined simultaneously. Now, almost two decades later, the various modifications that had been made as technology had advanced from the pendulum to the quartz crystal frequency standard were being incorporated in the new design.

A PZT is mounted rigidly in a vertical position, and its range of operation is limited to those stars which culminate close to the zenith in their apparent daily motion from east to west [Plate 36].[9] It is a reflex instrument. Light enters through an objective at the top to a mercury basin at the bottom, then is reflected back on itself and comes to a focus on a small photographic plate close to and just below the objective. As a consequence of the reflex characteristic, a PZT with a focal length of, say, fourteen feet requires little more than eight feet of head room [see Figure 1].

It is composed of three parts: a rotary, a main tube, and a mercury basin. The rotary contains the object glass, the carriage for the photographic plate, the lead screw, and the motor box. It takes its name from the fact that it rotates as a unit through precisely 180° after each twenty-second exposure. In order to secure point images, the plate is moved by a precise lead screw to follow the star's motion. Control for the lead screw is a synchronous motor geared to permit the screw to be turned in either direction. No matter which way the rotary is oriented, head east or head west, the plate may always be driven from west to east. Further discussion of the plate and its reduction will follow later.

McClenahan was most favourably impressed with what he saw in Washington. The US Naval Observatory Time Service generously made available a full set of drawings so that steps could be taken to have them modified to meet the Ottawa requirements. These were that the objective be 10 inches instead of 8, and the focal length 14 feet. The increase in the size of the objective from 8 to 10 inches followed the design of the Greenwich PZT because it was felt that it might be necessary to go to fainter stars. The increase in focal length over the PZT s in the United States (12 feet, 6 inches) and in Herstmonceux, England, (11 feet, 3 inches) provided for a plate scale of close to 48 seconds of arc per millimetre.

Vickers Ltd of Montreal formed all the castings, after first redrawing the Washington plans. The casting and figuring of the objective were assigned to Perkin-Elmer Company of Glenbrook, Connecticut. The other precision components associated with the control of the photographic plate, including the plate carriage, the lead screw, the 1000-cycle motor, and the sequence-control gear box, were made with slight modifications directly from the USNO drawings in the Dominion Observatory machine shop. J.P. Henderson suggested at the time that a Bodine 60-cycle motor, operating from controlled 60-cycle source, would do better than a custom designed 1000-cycle motor, and would be more readily

9 M.M. Thomson, 'The Ottawa Photographic Zenith Tube'

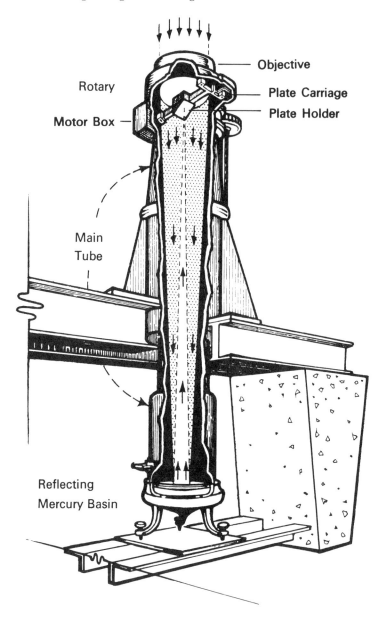

FIGURE 1 Ottawa Photographic Zenith Tube (PZT) schematic.
Courtesy Earth Physics Branch, EMR

replaced. It was only in 1958, two years after he retired, that the design was modified to use his proposal.

In order to have the PZT readily available for test and development, it was mounted in the transit room using two of the three piers originally designed for the Cooke transit. The first test exposures were made in mid-summer 1951, and by January 1952 it replaced the Cooke as the instrument for determining astronomical time at the Dominion Observatory.

Star positions for the PZT based on the FK3 catalogue were supplied by H.R. Morgan of the US Naval Observatory. At the end of the first year of observing, the PZT results were used to smooth the positions in both right ascension and declination. Meridian circle observations, necessary to improve the attachment of the catalogue to the FK3 system were made during 1950 to 1953. As evidence accumulated through the years, R.W. Tanner was very careful to apply all the observational material to each star in subsequent catalogue revisions.

Initially the Ottawa PZT catalogue consisted of nearly 160 stars. These were divided into twelve groups, each covering two hours of right ascension, and with an even distribution of stars north and south of the zenith in each group. Subsequently the catalogue was reduced to eighty stars which were divided into eight groups of three hours each. This meant that during the summer solstice, when seven hours only of night time observing was available at Ottawa, two full groups could still be observed. The winter solstice at the latitude of Ottawa, by contrast, provided fourteen hours of darkness. For the PZT, since observing was restricted to a half-degree cone centred on the zenith, exposure could commence at the end of civil twilight.

The moon never rises high enough to be overhead and hence to shine directly into the PZT at Ottawa. A moon shade had been designed initially, consisting of a sheet of plywood with a hole of about fourteen inches diameter suspended horizontally about three feet above the objective, but it was later abandoned. On a moonlit night a stray cloud passing overhead during an exposure, or even a light haze, could cause unwanted light to enter the PZT and fog the plate. Each exposure required twenty seconds, and each star four exposures, so that total exposure time for the night would be half an hour or more. City lights nearby caused no fogging. Up to eighty-three stars have been recorded on a single plate on a long winter night, involving nearly two hours of exposure time, without undue fogging or difficulty in identification.

At first a fast plate, Eastman 103-0, was used to be sure of recording the faintest star of the catalogue, which was between 9th and 10th magnitude. Subsequently, when the catalogue was restricted to stars whose magnitudes ranged from 5 to 9, tests showed that the slower emulsion Eastman 11-0 produced smaller, sharper images with less fogging and graininess, so it replaced the 103-0 plate.

An analysis of the various errors attendant on the PZT performance led to the expectation that the night-to-night scatter should be of the order of $\pm 0^s.006 = \pm 0''.06$. Instead it was nearly three times this value. This circumstance was principally blamed on the fact that the transit room was constructed of the same solid stone and masonry as the rest of the Observatory building, and in spite of fans to encourage good air circulation a temperature differential between inside and out of less than two degrees centigrade was seldom attained.

The big improvement occurred in 1960 with the removal of the PZT from the transit room to a properly designed observing hut made of asbestos siding attached to a metal framework. It had very low heat capacity. Furthermore, the whole roof was designed to roll back, so that the instrument was essentially exposed to the outside air. Satisfactory results were reported: 'In 1960 May, the PZT was moved from its temporary location in the transit room to a properly designed hut with a consequent drop in night error in α and δ [dispersion of the nightly values from the adopted smooth curve] from 16 milliseconds to 7 milliseconds, and from $0''.09$ to $0''.04$ respectively.' The precision originally expected had been achieved.

The location of the observing hut on the grassy slope in front of the observatory was only a temporary site, made available on a short-term basis by the Department of Agriculture during the period that the mirror transit circle (MTC) was being developed. Five years later when definite plans were initiated for its removal, the US Naval Observatory was in the midst of plans for the construction of a new PZT to be located in Pinto Indio, Argentina, on the same latitude as the PZT at Mount Stromlo, Australia. Considering the small additional cost of constructing two instead of one, arrangements were made for a second Canadian PZT to be made at Washington, incorporating improvements that had been developed both in Washington and in Ottawa. The new instrument, when finally completed, had an eight-inch instead of a ten-inch objective; but with all surfaces coated, it proved to be entirely adequate for the Ottawa PZT catalogue. So the older instrument, whose larger objective matched the Herstmonceux PZT, was established just south of Calgary, Alberta, on the same latitude as Herstmonceux, England, to form the first northern hemisphere pair of PZTs to use an entirely common star catalogue. Washington and Tokyo are close enough in latitude that they share a few stars in common. Subsequently a new instrument of eight-inch objective has replaced the older one at Calgary, and again the coating on each of the optical surfaces has made it capable of recording all the stars of the Herstmonceux catalogue. Also, the focal length of the new PZT was the same as the Herstmonceux one, so that the plate scales were identical. The scientific advantage of two PZT s on the same latitude and hence using the same star catalogue was that a relative shift of either instrument due to some geophysical reason would be readily detectable in spite of small errors in catalogue positions of the stars.

Attached to the lead screw which drives the photographic plate of the Ottawa and Calgary PZT s is a small cylinder with a slit which permits photoelectric, as opposed to mechanical, timing pulses to be recorded for each exposure to the tenth of a millisecond, a tenfold improvement over the transit telescope.

The shutter, when closed, is in the form of a loose-fitting cap which fits down over the objective but without coming in contact with it. The cap is divided into two halves with a small overlap, and each half is hinged outboard from the objective on an arm which is mounted on the platform and is quite independent of the PZT. The starting and stopping of the plate drive and the opening and closing of the shutter are carefully synchronized. Besides excluding unwanted light, the shutter, when closed, is an effective dew cap and dust cap. Occasionally when a sudden shower has overtaken the site, the shutter has also kept the rain from the object glass.

Automatic control of the PZT is provided by a 24-hour timer set to open the roof at the end of civil twilight at dusk and close it again at the beginning of civil twilight at dawn. If a sensor mounted above the roof line detects precipitation in the form of either rain or snow, the sensor flap and the roof will close. Twenty minutes later the flap will open, and if precipitation has ceased the roof will open and observing will resume; otherwise the roof will remain closed for another twenty minutes, and so on.

The other piece of automation is the program machine, designed by V.E. Hollinsworth in 1952, which controls the start and stop of each exposure [Plate 37]. It consists of one thousand feet of 35 mm film drawn from one reel by a synchronous Bodine motor and wound up on a second reel. On the way the film passes under four fingers, and perforations in the film allow the fingers to drop into holes in the bed plate, thus providing sufficient motion to close the appropriate contacts. Two contacts control the start and stop of the PZT observing sequence and the other two control the fan which draws air down through the tube at all times except during an observation. The air within the tube is thus prevented from becoming layered with injurious impact on the light beam.

An essential part of the program machine is the clock, whose sweep second hand is geared to the sprocket which draws the film so that any point on the film is represented by a time indication on the clock. With the motor declutched the film may be cranked forward or backward and the clock hands will follow exactly. The motor is driven synchronously by a 60-cycle sidereal frequency synthesized from the same crystal clock which controls the PZT plate drive. Each morning when the plate is removed, the program machine is turned back twenty-four hours and re-set for the following night. A thousand feet of film will hold more than 120 hours of information, so that there is a generous overlap.

The mercury basin of the PZT is important because of two functions which it performs: reflection and focus. As a reflecting surface it defines the vertical as determined by the force of gravity at the spot, and because it can be adjusted up or down it is used to focus the stellar images on the photographic plate. In order to make the mercury surface most effective, it is contained within a shallow dish with a sloping edge so that any wave on it will be shallow and readily damped by the sloping beach at the edge. Also, the dish is lined with copper, which readily forms an amalgam, so that the mercury tends to climb up the sloping edge rather than form a convex meniscus as it does in a glass tube. Finally, to minimize the effect of vibrations, the mercury dish is floated in a larger basin of mercury.

The scum which forms continuously as a consequence of the copper amalgam, dulls the mercury surface, and must be removed. A glass rod drawn across the surface is most effective. In a test dish lined with gold the mercury remained bright and clear for many nights, entirely free from scum, and subject only to dust accumulation. Gold and mercury have such an affinity for each other, however, that before long the thin lining had been absorbed, leaving the iron dish on which the mercury would not flow.

The late D.F. Stedman, who made many notable contributions in applied chemistry at the National Research Council, took a special interest in the problems of the mercury reflector of the PZT. It was under his direction that the copper lined dish was 'wetted,' using hydrochloric acid as a cleanser and adding to it a drop of mercury which spread as a thin film over the copper. He prepared the gold-lined dish, knowing that it was

24 / The Dominion Observatory, 1905 to 1970

25 / TOP LEFT R. Meldrum Stewart, director of the Time Service 1905–23, director of the Dominion Observatory 1923–46
Photo by Malak, 1943

26 / LEFT W.S. McClenahan, chief of the Time Service, marking the time on the drum chronograph prior to taking an observation with the transit instrument
National Film Board

27 / ABOVE The hoist for reversing the meridian circle telescope is stowed out of the way when not in use. The platform on top makes it a viewing stand for looking down through the telescope to determine the nadir with the help of the mercury basin. To the right on the pier is the collimator.

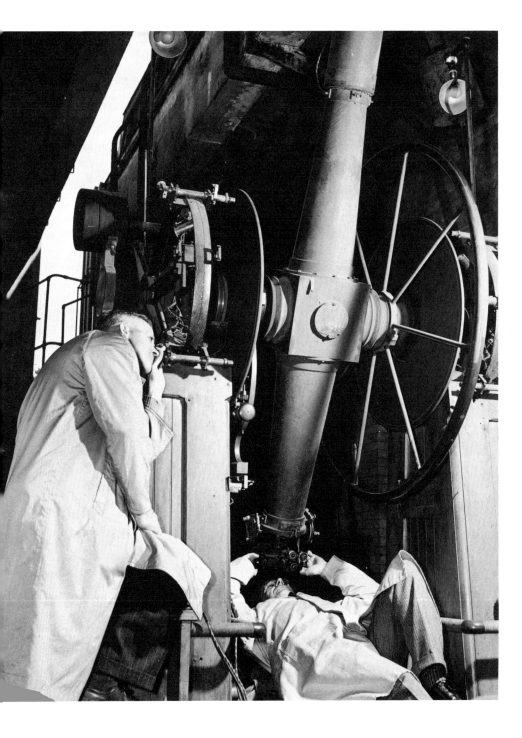

28 / LEFT The meridian circle transit telescope. R.J. McDiarmid, on the observing couch, has his hands on the micrometer wheels of the eyepiece. W.S. McClenahan is at one of the four microscopes used to record the instrument setting. Microscopes were later replaced by cameras.
Photo by Malak, 1943

29 / BELOW Sundial at Fort Providence, Northwest Territories, mounted on a transit pier of the 1921–2 survey. At latitude 61° 21′ north, the gnomon appears to be close to the vertical.

30 / C.C. Smith, chief of the Time Service at the Dominion Observatory, serves as timekeeper at the August 1932 solar eclipse camp established by the Observatory at St Alexis-des-Monts PQ.

31 / D.B. Nugent with his hand at the five-minute shaft of the time signal machine designed by R.M. Stewart
Photo by Malak, 1943

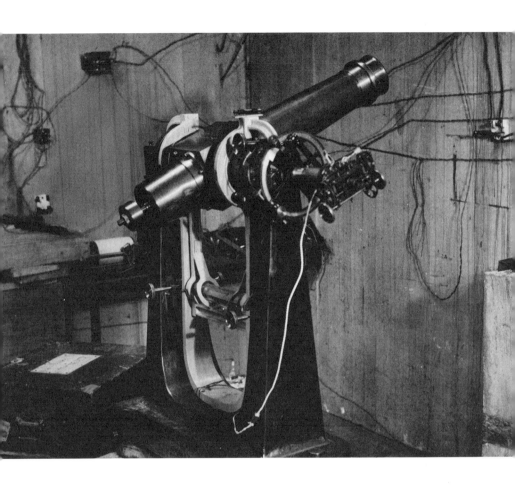

32 / The broken-type Cooke transit at the Brockton Point site, Vancouver, 1933, with a drum chronograph in the corner

33 / The author in an observing position at the broken-type Cooke transit in the transit room of the Dominion Observatory in 1948. The clock on the wall reads sidereal time.

34 / TOP LEFT The mirror transit telescope at the Dominion Observatory involved an optically flat mirror in a central position from which a star image was reflected into one of two horizontal telescopes located to the north and south. The observer is shown at the eye end of the north telescope in 1960.

35 / LEFT C.S. Beals, director of the Dominion Observatory 1946–64

36 / ABOVE The Ottawa photographic zenith tube in its hut with the roof partially open in 1962. The shutter is open, exposing the instrument to the sky. The hood at the bottom is raised, showing the mercury basin.

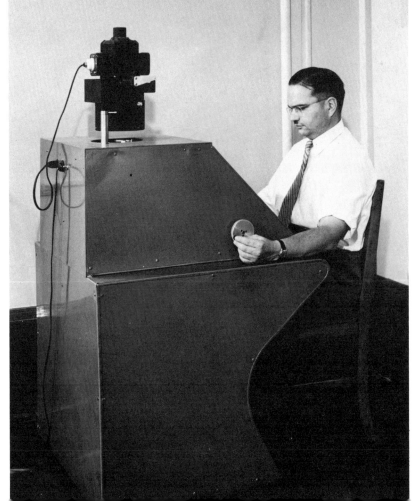

37 / TOP LEFT Front view of the PZT program machine designed by V.E. Hollinsworth in 1952. The punched tape is made of 35mm movie film. Sprocket and gear connection to the clock provides positive timing. Pilot lights indicate the various operations of the PZT.

38 / LEFT The PZT measuring engine designed and built at the Dominion Observatory in 1954, with L.G. Miller at the controls. It has since been adapted to on-line card punch, and a duplicate has been made for use at the Calgary PZT observatory.

39 / ABOVE The PZT measuring engine with star images projected onto the viewing screen. Hand wheels move the plate so that each star may be brought to the intersection of the lines on the screen.

40 / ABOVE Markowitz dual rate moon camera. The disc is the tilting filter which compensates for the motion of the moon and reduces its brightness to that of a star. The plate holder can also be driven at stellar speed if the telescope remains fixed.

41 / RIGHT Markowitz dual rate moon camera adapted to the fifteen-inch refractor of the Dominion Observatory. The telescope is driven at stellar speed, and the plate holder of the camera remains fixed.

⊙
-13° 5716
8·3 F5

⊙
-14° 5815
8·4 K0

⊙
-14° 5
4·C

⊙
-14° 5812
7·8 M3

⊙
-14° 5797
9·0 K0

42 / A picture of the new moon secured with a twenty-second exposure on the Markowitz moon camera. North is to the top. Stars have been accentuated for reproduction. Each star is identified with its BD Catalogue number, its magnitude, and its spectral type as listed in the Yale Zone Catalogue. Light from the moon has been intercepted by the tilting filter.

fragile, but from a desire to learn how much of an improvement a scum-free surface might be. This was followed by an unsuccessful attempt to obtain by electrolysis a thin film of another noble metal amalgam, palladium. But nothing has been found adequate to replace the copper-lined dish.

The focus of the PZT was determined initially by the use of a tape measure, followed by test exposures to find the sharpest image. An invar rod with a sharp point was then adjusted to preserve the distance. But a sensitive indicator for the focus was discovered by accident one night when half the shutter failed to operate, leaving only the west half of the object glass exposed. The fault was noted when the plate was removed. A decided offset in the clock correction for the night was blamed on the fault, but in order to be sure a special test was made the next night. A few stars were recorded with the west half of the shutter inactive, followed by a few more with the east half inactive. The difference in clock correction showed conclusively that the images were all being recorded beyond the true focal point, as shown schematically in Figure 2. Since then, periodic tests have been made using this method, and as a result the elevation of the mercury surface can be adjusted to the nearest millimetre.

Originally the PZT plates were measured using a Toepfer measuring engine, the principal drawback of which was the strain of adjusting to three optical systems using different microscopes. In the design of a new measuring engine certain definite objectives were sought [Plate 38]. As much as possible, visual microscopes would give way to projections, thus minimizing eye strain, and measurements in both co-ordinates had to be made accurately and easily. A seven-times enlargement of stellar images was considered adequate, though this was later increased to about eighteen times. The decision to use projection, eliminating the use of visual microscopes, arose from the success of previous developments using this principle at both Victoria and Ottawa.

The final machine, which I designed, with considerable input from C.S. Beals, is still in use in Ottawa, and a copy has been made for use in Calgary. The plate is illuminated and the image projected by an enlarging lens onto a ground glass viewing screen tilted at 45 degrees [Plate 39]. The fiducial lines are drawn with India ink directly onto the ground glass screen, thus avoiding parallax. The two ways are moved by precision millimeter screws belt-driven from two conveniently located hand wheels, one on each side of the machine. Geared to the lead screws are digitizers whose outputs are fed directly into an on-line card punch. The computations involved in the reduction of the PZT observations at Ottawa and

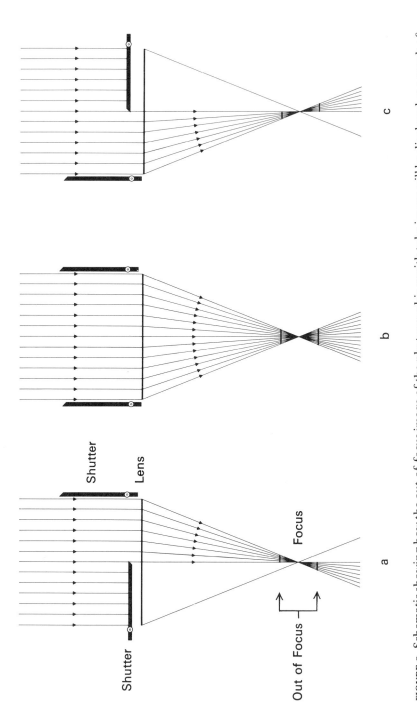

FIGURE 2 Schematic showing how the out-of-focus image of the photographic zenith tube image will be displaced as a result of half the shutter remaining closed. *a* west half closed; *b* both halves open; *c* east half closed.

Calgary have been adapted to data processing, and are available week by week for the international offices, namely the Bureau International de l'Heure in France and the International Polar Motion Service in Japan. The plate provides a permanent record for that date of the time that each star crossed the PZT meridian, together with its distance north or south of the zenith. The two values, time and zenith distance, are obtained by measuring the position of the four images of each star to the micron (thousandth of a millimetre) in a two co-ordinate measuring engine. The results, variation in the speed of earth rotation, and change in latitude due to polar wander, are of great importance to a better understanding of earth dynamics, beside yielding mean solar time.

In 1970 the Ottawa PZT was removed from its position on the lawn in front of the Dominion Observatory, and relocated about ten miles due west at the Quiet Site of the Department of National Defence. Its new position was measured by the Geodetic Survey as $49^s.323$ west of its former position. However the difference as determined by PZT observations is $49^s.199$, an apparent discrepancy of $0^s.124$ (3.8 m or 12.5 ft), which must be credited to change in the vertical caused by deflections of the force of gravity. Through such deflections the distance between two adjacent locations measured by astronomical methods can differ from the distance measured by the surveyor's tape along the ground. W.F. King, more than sixty years earlier, as we have noted, had pointed this out as the limitation to the astronomical determination of control points in mapmaking. In the Ottawa case, however, it was the astronomical value itself that was required.

Both the Ottawa and the Calgary PTZs are able to secure observations on about 60 per cent of the nights throughout the year. The average night-to-night variation in the results (or 'scatter') of the Ottawa observations is 3.5 milliseconds in time and $0''.04$ in latitude, a uniformity which places it to the forefront internationally. The Calgary record, perhaps because of the Chinook winds, is not quite so uniform, the night-to-night scatter being 6.5 milliseconds in time and $0''.04$ in latitude. But both instruments are making a valuable contribution to the geodynamics of earth rotation and polar wander.

During the eighteen months of the International Geophysical year 1957–8, in response to a request by N. Stoyko of the Bureau International de l'Heure, the broken Cooke transit was re-established on what had been the north collimator pier in order to secure a set of observations comparable to those taken before the introduction of the PZT.

In July 1957 Dr William Markowitz of the USNO provided a dual rate camera designed for the 15-inch telescope to measure the moon's position in the stars [Plates 40, 41]. In preparation, G.A. Brealey redesigned the eye end of the telescope to give a wider breech and a more sensitive method of focusing. A special camera was needed because the moon moves relatively rapidly through the stars, 0".55 per second, as opposed to the more leisurely motion of the sun, 0".04 per second. To take a picture of the moon and the stars simultaneously, a tilting filter in the camera was located close to the plate surface, its tilting motion computed to cause a displacement of the lunar image on the plate exactly equal to the apparent motion of the moon. At the same time the density of the filter was adequate to reduce the light from the moon to the equivalent of an average star. Another motor within the camera (hence the name 'dual rate') caused the plate to move with the stars. Since the Ottawa 15-inch was synchronously driven using a crystal-controlled 60-cycle frequency, the plate motor of the moon camera was disconnected. All of the plates were sent to the United States Naval Observatory for measurement and reduction after the star field had been identified [Plate 42]. The moon camera program continued at Ottawa for a few years after the International Geophysical Year before being finally suspended in 1963.

The use of the moon as an hour hand against the background of stars was made possible because of two programs, one relating to lunar motion and the other to the lunar profile. Electronic computation replaced desk computers, so that a new set of moon tables was made possible in strict accord with the intricate gravitational theory of the moon. Dirk Brouwer of Yale, G.M. Clemence of the us Nautical Almanac office, and W.J. Eckert of the Watson Computing Laboratory co-operated in this monumental task. The lunar profile is both rugged and changeable, the former due to the mountains and craters which cover much of the surface, the latter due to libration.[10] C.B. Watts of the us Naval Observatory made an extensive survey of lunar profiles, the results of which enabled the profile on any given date to be related to the same centre of gravity as the profile for any other date and ephemeris time to be calculated.

10 Libration: The moon keeps the same face towards the earth, so that the far side has been revealed only by recent lunar space ships. But because of an apparent wobble, caused by the fact that its speed of motion around the earth is not uniform, we see a little bit around one side and then a little bit around the other. Also, because the moon's equator is inclined 6^1/$_2$° to its orbit, we can see a bit over the north pole, then a bit over the south pole. Finally, when rising and setting, the moon is in a position which permits us to catch a glimpse over the edge. These three sources of libration allow 59 per cent of the moon's surface to be seen from the earth.

A drawback to the moon camera method was its inability to compensate for a slight shift of the lunar image during exposure. A stellar image, enlarged through poor seeing, can still be centred upon. The corresponding circumstance would be the ability to see the full orb of the moon. In ordinary light this is possible only during the brief interval of full moon, nearly half the lunar profile being obscured at all other times. So in every exposure of either ten or twenty seconds there is some uncertainty as to the exact position of the limb, and hence of the centre of gravity of the moon, that is, its position with respect to the stars.[11]

The advantage of the moon camera is that photographs are not limited to the meridian, but may be taken in any part of the sky. They are taken in pairs, the camera being reversed through exactly 180°. The time of an exposure is the instant that the tilting filter is exactly parallel to the photographic plate, and is determined from a set of contacts made by the filter during its rotation. Universal time is used to record the event. The position of the moon in its orbit is tabulated in ephemeris time, and hence appears to be in the wrong position both in right ascension and declination according to the time by the observing clock. It is this error, translated into a time interval, which is the difference between universal and ephemeris time.

The moon camera seemed to produce its best results in the hands of its inventor, William Markowitz of Washington, DC. It did not live up to expectations, and a few years after the conclusion of the IGY, the ten or so additional cameras, including the one at Ottawa, had ceased operations. Transit observations of the moon, using the transit circle with micrometer setting for both right ascension and declination, are obtained routinely by several observatories. The large number of lunar occultations observed by amateurs, with improved technique and accurate timing, is processed annually at the Nautical Almanac Office of Great Britain and afford further evidence of the reading of the ephemeris clock.

The PZTs at Ottawa and Calgary continue to provide mean solar time (UT1), which is required for all branches of earth satellite astronomy and surveying as well as general timekeeping. Through the years the consistency of the Ottawa PZT has been unsurpassed by any of the dozen or more

11 The limb of the moon is its visible edge. It is sharply defined because there is no shell of gas or air surrounding it as there is around the earth. A star may be watched as the moon approaches it, and then with a startling suddenness the star disappears behind the limb, and an occultation has occurred. Much of the moon's surface is rugged and mountainous. A star near the north or south end of the lunar path may be occulted by a mountain peak which sticks out on the limb and reappear for a moment on the other side.

instruments distributed around the world. At both Calgary and Ottawa are satellite doppler tracking stations which are a part of the global TRANET network for tracking earth satellites as a means of studying earth rotation, polar motion, and other branches of geodynamics. Radio has the advantage that observations can be made regardless of weather conditions, and while doppler observations surpass the PZT as a means of studying polar motion, they lack the precision of the PZT in earth rotation. It was estimated that the Calgary-Herstmonceux link would require twenty years to reveal a sensible shift due to continental drift, if indeed such a drift exists. Half that time has elapsed since the Calgary PZT was inaugurated, and there is every indication that the PZT as an optical instrument will continue to be useful for at least another decade. Ottawa and Calgary are linked by direct line, so that the results of doppler tracking of earth satellites and PZT observations can be made known with minimum delay.

5

From pendulums to masers: clocks of the Time Service

The new observatory building completed in 1905 provided ample space also for the non-astronomical functions for which the director, W.F. King, was responsible. R.M. Stewart was particularly pleased to have a clock room for the primary clocks, a time room for the secondary 'work' pendulums with all their relays and outgoing circuits [Plate 43], and office space for study and research.

The clock room was located in the very centre of the basement where there would be the least disturbance from traffic passing around the building. Double doors ensured thermal control. Uniform temperature was maintained by a thermostatically controlled heater, and a fan behind the heater kept the air of the room well mixed. Various improvements in the thermal control culminated in a Callendar Recorder in the time room with a platinum sensing device in the clock room. Control of the ambient temperature to about one-tenth of a degree centigrade was maintained.

Four piers were built, each on its own footing and independent of the walls or floor of the clock room, and on them the four primary timekeepers were mounted. They were arranged in pairs at the two ends of the room, a sidereal and mean time pendulum constituting a pair, and so disposed that the planes of oscillation of each pair of pendulums were at right angles to obviate any mutual effect.

The Howard clock, which had been purchased in 1900 for $500 and had served as the sidereal clock in the basement of the Supreme Court building, was moved to the new clock room to become the primary sidereal timekeeper. A Riefler pendulum served as standby. These two clocks were adjusted as nearly as possible to zero rate and then left

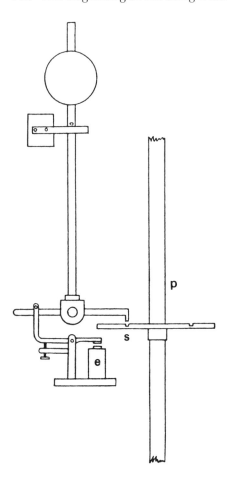

FIGURE 3 Pendulum correction. The large pendulum *p*, of which only a portion is shown, may be advanced or retarded by one of two auxiliary pendulums, one of which is shown on the left, which may be attached to it by activating the electromagnet *e*, allowing the tooth to drop into the slot *s*. If the auxiliary pendulum is supported above its centre of mass, it will advance the main pendulum, if below, retard it.

undisturbed in the expectation that any residual rate, as determined by star transits, would remain small and uniform.

Each morning the practice was maintained of computing the correct value of mean time from the best known value of sidereal time. And a switch was thrown to the left or right on the control panel of the time room to advance or retard the mean time primary by the required amount, just as it had been done from the office in the Thistle Building in the early days. Now, however, a new and improved method of advance and retard was employed [see Figure 3]. Two small pendulums were mounted, one on each side of the mean time primary, close enough that one or other could be attached to the main pendulum. One was suspended in the normal manner with its centre of mass below the point of support; its much shorter natural period caused the main pendulum to swing a little faster. The other small pendulum was suspended below its centre of mass, and its attachment caused the main one to swing more slowly. The amount of gain or loss amounted to about 0.01 second per second.[1]

The Cornu method of holding secondary pendulums in synchronism [see p. 71, n. 14] had been replaced by the Riefler system. R.M. Stewart described the latter as follows: 'In each of the synchronized clocks was a 'synchronizing' electromagnet situated below and to one side of the end of the pendulum rod; attached to the pendulum was a small armature of soft iron, which would be attracted to the electromagnet whenever current flowed through the latter. A current controlled by the primary clock was arranged to flow through the synchronized magnet every alternate second; thus the pendulum of the synchronized clock was forced to swing in synchronism with that of the primary.' The system, however, could not withstand an interruption in the control line from the observatory to the secondary clocks. The latter would generally not retain a steady rate independently, but would soon get out of phase. When the line was restored the secondary clock would invariably stop, and the dials dependent upon the clock would also stop. Where the line was free of interruption, of course, such as in the confines of the observatory, the system worked satisfactorily for many years. It is the method by which the Howard, which is today mounted in the Physics building of the National Research Council in Ottawa and adjusted to a mean time rate, is held in step with the cesium standard. But the relatively long lines from the observatory to the clocks in the government buildings were frequently interrupted, and so were not suitable for the Riefler method.

1 R.M. Stewart, 'The time service at the Dominion Observatory'

A unique system for holding a clock in synchronism with its primary was introduced by Stewart himself in 1913.[2] It consisted of a tray mounted near the upper end of the pendulum, and two small weights which could be lowered onto or held free from the tray [Plate 44]. When the weights rested on the tray, the centre of gravity of the pendulum was raised, the period of swing was reduced, and the clock would gain. Conversely, when the weights were held free the clock would run slow. Should the power to the control magnets fail, one weight was held free and the other rode the tray, and in this situation the pendulum was regulated as nearly as possible to zero rate. A synchronizing pulse from the master pendulum once a minute determined whether the weights would ride or be held free for the ensuing minute. The secondary master, whose rate could vary by six or eight seconds a day due to thermal fluctuations, could be held to the time of the primary to within about a hundredth of a second. The synchronization line to the clocks in the government buildings in downtown Ottawa, described above, was activated once an hour, and this maintained all the minute dials in the government offices within a few hundredths of a second. When the dials advanced only once per minute, or once per half minute, this precision was quite satisfactory.

In November 1905, a new model of the Riefler clock had been acquired and installed as the primary sidereal clock of the observatory. Numbered R75, it incorporated the new Riefler concept of a nickel-steel pendulum, a free escapement, and an electrical self-winding mechanism, all enclosed in an airtight glass cylinder. The self-winding mechanism was in the form of a gravity arm centred on the axis of the escapement wheel and with a ratchet attachment to the teeth of the wheel. It rode down with the turning of the wheel till it touched an electrical contact, whereupon it was flipped to its upper position. Pressure inside the cylinder was maintained at 725 mm of mercury, slightly below barometric pressure. A few volts of battery power could keep the clock in operation for many months, while a second pair of wires through the vacuum seal operated a relay from the pendulum contacts. The Howard clock was adjusted to mean time and subsequently became the mean time primary. Though it marked a great advance in reliability of performance, R75 was apparently sensitive to small thermal fluctuations, and R.M. Stewart went to great pains to provide a better environment, such as a special wooden enclosure with thermostatically controlled air flow.

In 1922 a second Riefler, R412, was ordered and placed in service. The

2 R.M. Stewart, 'A new form of clock synchronism'

two complemented each other and brought to light the need of an improved environment which would be more spacious, free from the effects of vibration from motors, and capable of satisfactory temperature regulation. A third precision pendulum was needed as well, because with only two it was not easy to determine which was developing a faulty rate.

In March 1923, a discrepancy showed that something was wrong with one of the two Rieflers. Failing a third precision pendulum, a Nardin sidereal chronometer was used on an all-night test, which revealed that the new clock, R412, was irregular.[3] A day later R412 stopped altogether because of a tight bearing on the hour-hand wheel, and R75 became temporarily the primary standard. Two years later R75 itself stopped from a failure in the winding mechanism. Each time it was necessary to break the vacuum seal for repairs and later to restore it.

The new clock vault was built 'under the north lawn of the observatory ... The outside dimensions of the vault were approximately 15 feet by 19 feet, with a height of 11 feet, the top of the roof being two feet below the surface of the ground. The walls and ceiling are double with a 4-inch air space. A 3-foot hall divides the vault into two parts, and each of these parts is divided into two rooms. Massive concrete piers, separated from the floor, are placed to support the clocks, and on each pier a half inch steel plate is bolted to give a firm and even surface for the holding of the clock brackets ... The temperature control in the hall is by bimetallic thermostat and in each room is a petroleum-mercury thermostat. The heating unit in the hall is a 660-watt heater and the rooms are heated by carbon lamps.'[4] The temperature of the vault was maintained at 25°c, which was higher than the ground temperature during the heat of summer. Fans kept the air circulating in each of the four rooms and the hall. Tile around the vault drained into a nearby collector pipe, which occasionally became clogged from tree roots, subjecting the vault to the inconvenience of several inches of water on the floor. The vault was completed in 1927; R75 was installed in it in 1928 and R412 the following year.[5]

A Shortt synchronome clock, master and slave, numbered s29, was obtained in January 1930. The master, or free pendulum, was installed in the vault [Plate 45] while the slave was mounted in the time room. This remarkable clock [Figure 4] incorporated a further advance in that no work was required of the free pendulum. It was sealed in a container that

3 J.P. Henderson diary.
4 C.C. Smith, 'Activities of the Astronomical Branch (history)'
5 C.C. Smith, 'Activities of the Astronomical Branch (history)'

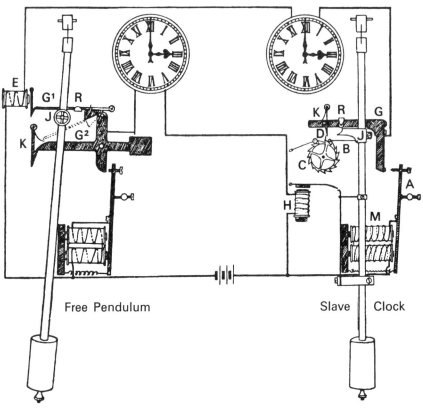

Free Pendulum   Slave | Clock

FIGURE 4 s29 free pendulum and slave clock. The slave does all the work. The arm
B, attached to the slave, advances the 15-toothed wheel *C* one tooth at a time. A pin
D on *C* releases catch *K* each half minute. The brass arm, pivoted at *G*, moves down
so that the roller *R* rolls down the slope of *J*, giving the slave the impulse which
keeps it swinging. The vertical member of the brass arm engages contact *A* and
energizes *M*, which restores the brass arm to catch *K*. By another route, contact *A*
energizes *E*, releasing gravity arm G^1, causing *R* to engage roller *J* on the free
pendulum. At the bottom of its run, G^1 releases G^2, which drops and closes the
contact on the tongue of the relay beside the free pendulum, restoring both G^2
and G^1 to the ready. The same contact energizes *H* as the slave is swinging to the
left. If the slave is late, the relay arm of *H* engages the vertical leaf spring clamped
to the slave pendulum, whereby the slave is given an accelerating impulse. If the
slave is not late enough, the leaf spring will pass beneath the arm of relay *H*,
resulting in a miss. The slave is rated to run slow with respect to the free pendulum
and normally at alternate half minute impulses the arm of relay *H* will engage the
leaf spring, resulting in an accelerating impulse. [F. Hope-Jones, *Electrical
Timekeeping*, 165; reproduced by permission of NAG Press Ltd., London]

was evacuated to about 32 mm of mercury. A slight adjustment in the vacuum could be made to adjust the rate. Every thirty seconds it received an impulse triggered by the slave's own half-minute impulse which restored the arc for another thirty seconds of free swinging. The impulse was in the form of a gravity arm carrying a jewel which rode down the periphery of a wheel attached to the pendulum. When the arm dropped to the bottom of its travel, it released a more massive lever which eased it back up and the lever was itself restored to the ready when it closed an electric circuit. The current from this pulse passed around the coils of an electromagnet adjacent to the slave, and the armature could engage a long leaf spring carried by the slave if it dropped sufficiently behind. The slave was rated to lose just enough that the armature would engage the leaf spring on alternate half-minute impulses, and when it missed engagement it would merely give a harmless tap to the top of the leaf spring. Thus the slave was able to reproduce the excellent performance of the free pendulum.

A brief period of testing demonstrated the superior performance of s29. It thereupon became the primary clock, and the two Rieflers were reduced to the role of secondary timekeepers. Had it not been for the financial stringency precipitated by the depression, an order would have been placed immediately for a second Shortt synchronome. The success of this clock inspired R.M. Stewart to develop his time machine, which mechanically produced mean solar time from a sidereal input; in so doing he tacitly pinned his faith on the pendulum as the ultimate instrument for timekeeping.

The ratio between mean solar time and mean sidereal time is one of the best known values in astronomy. It is based on the fact that during the ordinary or tropical year of 365.2422 days, an extra day has gradually accumulated with respect to the rising and setting of the fixed stars, with the result that there are 366.2422 sidereal days. Every rotation of the earth with respect to the mean sun results in a little more than one rotation with respect to the fixed stars. It is a well observed fact that the same star rises earlier night after night. A mean solar day equals 1.002737909 sidereal days. Because of this ratio it is a relatively easy exercise to determine the mean solar time if one knows the date and the sidereal time, and vice versa one can compute sidereal time from mean time.

Stewart may not have been the first person to ponder the possibility of a mechanism which would automatically deliver mean solar time from a sidereal source. The problem was to produce 365.2422 mean solar seconds. He was successful. July 1937 marked the first test of his new time machine, which was placed in operation the following year.

He started with a mean solar half-second pendulum, the kind that swings once a second instead of the standard pendulum which swings once in two seconds. By combining gear ratios of 79:130, 70:123, and 57:120, he arrived at a value of 60:365.2422. Hence from a shaft which rotated once in sixty seconds of mean solar time he caused another shaft to rotate once in 365.2422 seconds. During this interval the sidereal pendulum would produce 366.2422 seconds. Now, to compare the shaft with the sidereal clock, two relays were used, one controlled by the half-seconds pendulum, the other by the sidereal pendulum. They would automatically come into synchronism in a little over six minutes, and when they did, they tripped a synchronizing mechanism. It should be noted that the rate of the sidereal clock, s29, was included in the gear assembly, so that the output from the half-seconds pendulum represented the sidereal interval as indicated by s29. If, however, the mean time pendulum were not correct, but were running slow, the shaft would not have completed its rotation in 366.2422 sidereal seconds. On the other hand the two relays would come into synchronism earlier than usual because the sidereal relay would overtake the time machine relay in a shorter time. Thus the shaft being slow, while the relay synchronism came early, accentuated the fact that the time machine pendulum was running slow and needed to be advanced. A similar situation in reverse would occur when the little pendulum had to be retarded to get back in step with the master sidereal, s29.

The indicator on the time machine was a figure-of-eight cam on the output of the 365.2422 shaft. Adjacent to it was the electromagnet which was activated when the two relays came into synchronism. An arm from the electromagnet forced the figure-of-eight cam into the horizontal position, bringing it back if it had over run, or advancing it if it were lagging. The same impulse unlatched a contact wheel and through a differential gear attachment caused it to rotate forward or back an amount equivalent to the error in the position of the figure-of-eight cam. This error was then applied as a correction to the half-second pendulum which controlled the escapement of the time machine.

The half-seconds pendulum was mounted on the wall directly above the machine [Plate 31]. It was only about one-quarter the length of a seconds pendulum, and so required little space. On an extension to the pendulum above the centre of support was a cone-shaped cup, into which was lowered a piece of brass when the sensing mechanism indicated that the time machine was in advance and had to be retarded. The advance was accomplished by a long leaf spring extending upward from the pendulum bob, which was engaged by the extended armature of an electromagnet and depressed as the swing continued to the right (as in the s29 slave; see Figure 4). On the return of the pendulum to the left the spring was released. The spring was depressed on successive swings till the proper advance had been achieved.

Even during the interval that a correction was being applied, the length of the second delivered by the time machine differed from a true mean time second by less than a thousandth, and the total correction seldom exceeded one hundredth of a second.

An arm extending to the left from the top of the pendulum bob pushed a toggle switch back and forth as the pendulum passed through the centre of its swing in each direction. The toggle switch gave the impulse to the escapement relay of the time machine.

The time machine was weight-driven, and the weight was wound automatically. In keeping with the latest design of pendulum clock, the act of winding increased slightly the torque on the train of gears. The weight of some three hundred pounds was applied to the slow-moving end, and the winding drum had a capacity of about a week. As one might expect, the heavy duty gears gave way to progressively lighter ones as the drive progressed from the slow motion of one revolution per day to the more rapid one revolution per minute.

The output of the time machine was contained on five shafts on which were notched discs. Fingers which rode on the discs and dropped into the notches made the required contacts. The five shafts had rotational periods of once a day, once an hour, once every five minutes, and two at once a minute. The slower moving discs acted as gates for signals from the faster moving ones. For instance a disc on the day shaft provided the opening once each day for the signal to the CN telegraph office. The signal awaited an opening provided by a disc on the hour shaft, and the actual signal was given by two carefully coded discs on one of the minute shafts. A similar process was followed for the signal to the CP telegraph an hour later. A synchronizing pulse to the secondary control pendulums in the downtown government buildings went out each hour, and involved only the hour and the minute shafts.

Seconds pulses for CHU, the radio transmitter, were initiated by the half-second pendulum itself in order to avoid the slight eccentricity which was almost impossible to avoid when mounting a disc on a shaft. The omissions (see below, p. 154), which included the 29th pulse and a coded sequence at the end of each minute of a five-minute cycle, involved a disc on the five-minute shaft. Also, the long dash on the hour, followed by the call sign CHU, delivered in Morse code twice in the first minute of the hour, involved a coded disc on the hour shaft as well as a minute disc with the Morse symbols.

A 24-hour dial with hour hand was mounted on the end of the day shaft, a 60-minute dial on the hour shaft, and a 60-second dial on the end of one of the minute shafts, so that at a glance one could read the time. The control was the escapement relay, operated as a uniform off-on impulse directly from the pendulum toggle switch.

It is not known for sure when R.M. Stewart commenced working on the design, but it is estimated that its construction monopolized an accumulated total of two years of machine shop time. The result was an ingenious combination of gears, and a monument to the skill of the machinist, L.P. Christenson, and his assistant, A. Bird. So far as is known, Stewart's time machine was unique. Nor was it copied, because the quartz crystal clock was just making its appearance.

There were two time machines, one always at the ready in case the other required servicing. Each was mounted on a heavy cast-iron base, which in turn was mounted on an oak bench. The two benches occupied adjacent walls in one corner of the time room. Glass dust covers protected both time machines and both pendulums. The cabinet space under each bench contained relays, wiring terminals, and the winding motor. Each machine was independently synchronized with respect to s29, and there was the possibility of one drifting away from the other. To warn that such might be happening, they were interconnected to an alarm bell which sounded if they differed in output by more than 0.015 second.

The time machines were set in operation in 1938, and for more than a decade they served their intended purpose. Each was capable of controlling the complete mean time requirements within the observatory, such as seconds dials, minute dials, the tower clock in the dome, and shutters for the seismic recorders. In addition they controlled all the mean time output, including the synchronization pulse for the government buildings, the continuous signals for CHU, the individually coded signals for each of the two telegraph companies, the noon signal for the CBC, and the special seasonal signals relayed to the Monitoring Station for the benefit of survey parties, and to the Department of National Defence for navigation on the North Atlantic.

V.E. Hollinsworth was responsible for wiring up the circuits and adjusting the discs on their shafts. His short stature enabled him to get right inside the cabinets under the time machines to work on the wiring. His engineering skill, together with the inventive genius of J.P. Henderson, was in no small way responsible for promptly bringing the time machines into successful operation.

The time machine system had a number of defects. The daily rate of s29, losing a little more than a tenth of a second a day, had to be accounted for mechanically, necessitating frequent adjustments. Occasionally, when s29 was at its own half-minute correction phase, the correction to the half-second pendulums could be exaggerated, amounting to as much as 0.01 second. The half-second pendulum may have had a variable rate because its protective case was merely dust proof and did not provide temperature or pressure control. Most noticeably, each machine operated with an escapement, and the consequent banging, starting and stopping, was neither good for the mechanism nor appreciated by the personnel obliged to work in the same room. Maintenance also could be tricky. Once, when the contacts of the fingers of the time machine were being serviced the escapement ratchet was accidentally lifted. Immediately the gear train started to accelerate out of control. The contact fingers danced on their discs like the fingers of a pianist until the heavy drive weight came to an abrupt stop at the bottom of its run. Two days were required to restore all the discs to their proper orientations.

On the other hand the pendulum timekeepers would give long, uninterrupted service which, even in the late 1940s, crystal timekeepers had not yet matched. Furthermore, the time machines were perfectly adequate from the standpoint of accuracy. In the words of a 1948 assessment of their performance:

The accuracy of the time signal machine is only slightly less than that of s29. And since the variations in the seconds beats of s29 from hour to hour are normally less than 0.01 second, the short period fluctuations in the time signal machine seldom exceed 0.01 second. From day to day the rate of s29 varied slowly between 0.13 and 0.16 second per day during 1947 but the overall rate was about 0.148 second. The short period fluctions of 0.01 second or less are of too small an order to be detected by surveyors and other government or private field parties, though the errors will be incorporated in their results. Nor are they of concern to the average Canadian who hears the CBC signal at one o'clock eastern time each day. They are, however, too large for present day laboratory measurements unless the measurements can be made over an extended period of time.

An opportunity to try a 60-cycle crystal-controlled motor on one of the time machines came in 1949 when s29 was accidentally stopped. For a while the time room enjoyed a delightful freedom from the thumping of one of the time machines. The circuitry had been available and tested, but lacked official approval. Now the superior short-term precision of the crystal oscillator could be demonstrated. But two months later a relay failed, the time machine drifted in error a matter of eleven seconds, and the CBC, with the same error, was embarrassed. An entry in the diary of J.P. Henderson for 29 November 1949 reads: 'Time Machine #2 put back on s29. Now both machines are on s29 – so many failures of crystal – all parts of relays have given troubles.' And yet the crystals had one important advantage: with pendulum clocks hours or days are required to determine rates; with crystal equipment accurate rate comparisons may be made in a few seconds.

R.M. Stewart never lost his faith in the pendulum as the ideal oscillator. With the assistance of Hollinsworth he pursued his research, even after his retirement in 1946, applying the principles of thermal control, thermal compensation, and photoelectric pick-off. But the pendulum had already been superseded by the quartz crystal frequency standard at Greenwich and Washington, and by 1951 it happened at Ottawa.

In October 1951 the arrival of two Muirhead timing devices [Plate 50] rendered the time signal machines obsolete except for that portion which generated the coded signals for the CN and CP telegraphs. They were converted to synchronous drive, and for a while operated in parallel with the Muirheads but later were dismantled. In their place two compact time signal machines were constructed, using only what shafts and discs and contact arms were needed to generate the coded time signals. At the time of writing they were still operating and the information was still being fed to the CN telegraph office, though the line to the CP telegraph was discontinued when the move was made from the Observatory to NRC in 1970.

The railway time signals were designed to be radiated across the two networks for two minutes each day, the CN from 10.58 to 11.00 A.M., and the CP from 11.54 to 11.56 A.M. In the first minute of the CN code, the relay was held on for one second and off the next, except that the 58th was omitted in order to emphasize the zero of the next minute. During the second minute the tempo was doubled, with the relay being closed for a half second and opened for a half second with the 58th and 59th omitted. In the CP code, the first minute had a short pulse at the beginning of the even seconds, and the second minute had two short pulses spaced half a second apart at the beginning of each even second. Again the 58th was omitted.

For test purposes, the signals were imposed onto the local loop twenty minutes or more ahead of time. The local CN office would then hear the familiar pulses at 10.38 to 10.40 A.M., 10.43 to 10.45 A.M., and so on. Similarly the CP would hear their pulses at 11.34 to 11.36 A.M. and at successive five minute intervals. In addition the c.p. line received T-I-M-E in Morse code just before 11.54 A.M.

The two Muirhead signal devices were controlled by 1000-hertz synchronous motors which were manually brought up to speed by a removable hand crank. They were continuously phasable so that the output contacts could be adjusted to the nearest millisecond. In addition to the normal signals required for the time service with an omission on the 29th second and 56th to 59th seconds of each minute, there was provision in the Muirhead machines for a lengthening of the minute dash from one-third to one-half of a second and of the dash on the hour to one second. CHU-CANADA-CHU was sent in Morse code twice during the first minute of each hour in place of CHU-CHU which appeared but once on the earlier time machines. Provision was also made for the transmission of sixty-one beats to the minute, following the method employed on some of the European transmissions, but it was never used.

THE QUARTZ CRYSTAL FREQUENCY STANDARD

In 1929, observatory personnel were drawn into a discussion with Department of Marine and Fisheries officials concerning electronic or electromechanical frequency standards. The minister of Marine and Fisheries, by virtue of the Radiotelegraph Act, 1913, was authorized to make regulations and prescribe conditions and restrictions to which a licensee was subject.[6] J.W. Bain, of the Radio Branch, said that, to check the tuning of their transmitters, the Branch was intending to purchase an absolute frequency standard from the National Physical Laboratories in Teddington, England. The output would be in the form of a small motor whose shaft turned at one revolution per second (rps). Bain asked

6 The regulations at that time conformed to international conventions, such as the 1912 London Convention. Article 4, section 2 of the 1927 International Radiotelegraph convention of Washington stated, 'Waves emitted by a station must be maintained upon the authorized frequency as exactly as the state of the art permits, and the radiation must also be as free as practicable from all emissions not essential to the type of communication carried on.' G.W. Richardson, 'A survey of Canadian broadcasting legislation,' *Can. Bar Rev.*, February 1937. The various regulations governing the administration of radio communication in Canada are now administered by the Department of Communication.

whether the observatory would have a chronograph and clock to determine accurately the interval of this output. Henderson's reply was yes, if a telephone line and proper terminal equipment were available.

Ed Davey, a wireless operator associated with the radio monitoring operation of the Marine and Fisheries Department from its inception, recorded in 1967 his reminiscences of these early days,

The first frequency standard operated by the old Radio Branch of the Marine and Fisheries Department that I have recollections of seeing in approximately 1932 was a Sullivan Fork, operating at 1000 cps [cycles per second, in today's notation called 1000 hertz]. The temperature control system for the fork chamber depended on the expansion of toluol confined in a fairly large glass tubular helix. After several 'flare-ups,' the system was condemned as being rather crude and a decided fire hazard ... This project was the responsibility of ... J.W. Bain ... The purpose of the Sullivan Fork was not for the measurement of remote signals, but rather for the calibration of wave meters which were in field offices for the adjustment of transmitters ashore and afloat, a number of which were of the 'spark' variety. Some time later, perhaps a year or so, Mr Bain was able to purchase the first General Radio Primary Standard of Frequency developed by J.K. Clapp of General Radio.

Initially the new standard, together with three Canadian Marconi TRF [tuned radio frequency] receivers, was installed in the test room on Wellington Street. Because of interference caused by the four-thousand-watt transmitter (call sign VAA) on the floor above and by the streetcars outside, the monitoring equipment was moved to the Booth Farmhouse on the Experimental Farm.

In 1934 a telephone line was installed between the Booth Farmhouse and the observatory. A frequency of 1000 hertz from the frequency standard operated a synchronometer which Mr Bain made available for the time room. The synchronometer contained a vertically mounted phonic motor with a shaft from the armature which had to be spun by hand to get it started. It turned at 10 rps and controlled both a clock dial and a 1 rps contact arm. The contact arm tripped a contact on a graduated drum which could be rotated and whose position could be read to a tenth of a division (a thousandth of a second). It was used to indicate the performance of the crystal frequency standard of the monitoring station with respect to the observatory primary clock. Hence it became known as 'Cm', meaning 'crystal clock monitoring station.'

'It wasn't long' Ed Davey continued, 'before the Time Room observed that the rate of our crystal clock was as good as, and sometimes better than their own time standards. This in turn led to the Time Room forsaking the Riefler and Shortt pendulums in favour of a crystal controlled time base.'

In 1932 a second body was established by Parliament with regulatory powers in the radio frequency field, the Canadian Radio Broadcasting Commission (CRBC). The activities of the CRBC were restricted to the broadcast band, 500 to 1500 kilohertz (kHz). The regulations required that each broadcast station should be so operated that the frequency would be maintained between the limits of 50 hertz above and 50 hertz below the assigned frequency. A three-component General Radio crystal frequency standard was purchased by the CRBC, and space was provided for it by the National Research Council on Sussex Drive in Ottawa.[7] The principal crystal was a quartz bar designed to operate at 50 kHz, while the other two quartz bars were cut to oscillate at frequencies of a few hertz above and below. Uniform beats between the principal and the two offset frequencies resulted in an elegant method of internal monitoring. But the inherent characteristic of quartz oscillators to drift in frequency as they aged required an absolute or fundamental frequency reference. At that time the astronomical pendulum, which was accurately monitored against earth rotation, was the only such reference. Accordingly, Keith A. MacKinnon, engineer at the CRBC monitoring room, installed a telephone line to the Dominion Observatory with which he was able to introduce the half-minute pulses of the free pendulum of s29 to his monitoring chart. It was necessary to allow for the rate of the pendulum as revealed by star observations and, in addition, to allow for the difference between sidereal and mean time. MacKinnon also monitored the time signals of the US Naval Observatory and the standard frequency transmissions on 5, 10, and 15 megahertz from wwv, the transmitter of the National Bureau of Standards, Washington, making a total of five external checks on his primary crystal. The feedback through the CRBC to the observatory from both these US monitoring sources proved to be very helpful in observing the performance of the observatory clocks.

Canadians were made aware of the CRBC frequency standard by the 'Oh Canada' time signal heard at 10 P.M. each night over the CRBC network, the last note of the bar marking zero of the hour. Ed Davey claimed that Alphonse Ouimet, then an engineer at Sussex Drive, later to become head of the Canadian Broadcasting Commission (CBC), was responsible for this contribution.

In 1926 J.P. Henderson of the observatory had built a vacuum tube

7 MacKinnon. 'Carrier frequency control'

oscillator to resonate at 10 hertz, and in order to use this frequency as a scale on the chronograph he adapted a Baldwin earphone with a syphon pen as a high speed recorder. Ten years later he was working with a 1000-hertz fork from which he counted down to 200 hertz and to 50 hertz by multivibrators. A small amount of power at 50 hertz was used to hold a special direct current motor in synchronism, and hence to drive a drum chronograph at 1 rps, and by gear reduction at 1 rpm. The greatly improved stability of the 1000 hertz from the crystal-controlled Cm of the Monitoring Station caused him to abandon the tuning fork, and also to add a 10-hertz multivibrator stage. The Baldwin phone and syphon pen responded readily to both the 10- and 50-hertz frequencies. It was therefore a useful tool in examining the short-period variations of the Riefler and s29 pendulums, and in measuring the lags of relays associated with the clocks and with the transit instruments. The contacts within the clocks, and the commutator contact associated with the micrometer of the transit instrument, were capable of light duty only. The heavy-duty operation of external circuits, such as the printing hammer of the chronograph or the impulse to the downtown time service, had to be handled by a relay.

Relay lag, the length of time required for the armature of the relay to respond to the impulse given to the coils, was readily measured by recording a series of clock beats, some with the relay in the circuit and some with the relay bypassed. One such measurement made in 1931 showed that the relay lag of the meridian circle transit micrometer was 0.025 second. When out of adjustment due to pitted points, excessive gap, improper spring tension, or some other reason, the lag generally increased, and had been observed to have three or four times its normal value. In spite of preventive maintenance, relays continued to go out of adjustment.

Once when C.C. Smith was measuring the relay lag from the transit instrument he asked Henderson if he could devise a relay with no lag. Henderson experimented with electronic devices in which higher voltages and sharper rise times could be used. In 1933 a 1200-ohm relay on the output of an old peanut tube recorder had a lag of 0.013 second. The same relay on the output of a more modern recorder using 227 tubes had a lag of about 0.008 second. A more sensitive relay of only 300 ohms resistance had a lag of only about a millisecond. Experiments were conducted with neon and argon gas tubes which, when fired with an input contact, would discharge a condenser storage through an output circuit such as the coils of a relay. More use was actually made of the flash from the gas tube to illuminate a sectored disc on the end of the 10 rps shaft of the synchronometer whereby Henderson was able to examine photographically the behaviour of clock beats. The contact of the synchronometer drum was shown to have an eccentricity of several milliseconds.

The thyratron, a three-element gas-filled tube of conventional design, was sensitive enough that it could be tripped directly from a telescope or clock contact and instantly exhibit the arc type of voltage drop through the plate circuit so that it offered very little impedance to the external circuit. It was well adapted to the heavy surge required by the hammer of the printing chronograph, since its input could be either a make or a break circuit.

Another electronic relay of less conventional design was contrived by Henderson using the ordinary audio amplifier tube. Having the grid held in bias through a high impedance network enabled him to provide control of the output by a high impedance low current contact to the grid such as with finger tips. He applied this to the pendulum of one of the time signal machines, the contact with the pendulum being a fine wire (number 40) against which it brushed when it swung back and forth through dead centre. The mechanical energy exacted from the pendulum was far less than was involved in the usual type of contact. The extremely small amount of current drawn by the grid of the tube caused little or no pitting or burning or sparking at the point of contact with the pendulum.

In 1938 Henderson also developed the inductive impulse method for the instantaneous operation of relays. The method used the break of a contact to release the stored up energy of an 8- to 10-henry inductance which had been charged by a direct current of about eight volts. An ordinary 110-volt neon bulb, with its base resistor removed, effectively blocked the low voltage from the external circuit, but presented no barrier to the inductive surge when the contact was opened. Several relays on the output of the time signal machines were operated in this manner.

The CRBC as a regulatory authority came to an end in December 1936. Radio monitoring was then consolidated under the authority of the newly formed Department of Transport. A year and a half later the frequency standard established at the CRBC by MacKinnon was transferred to Bain and Davey at the monitoring station of DOT, augmenting their equipment, which now included a Marconi as well as a General Radio standard. Part of the delay in the transfer was due to the continuing need for standard frequencies within the laboratories of NRC, and it took a while for the priorities between NRC and DOT to be settled.

When D.W.R. McKinley, a recent graduate of the University of Toronto, arrived at NRC late in 1938, one of his first chores was to provide a standard frequency throughout the labs, as had been done by MacKinnon. Initially he arranged for the transmission by telephone line of 1000-Hz frequency from the monitoring station via the observatory. Later he assembled the material for his own generator. By the end of 1938 he was able to synchronize his standard with wwv to one beat in ten seconds on the 5 megahertz transmission, which meant about 2 parts in 10^8. Since Bain at DOT also took regular readings of the US Naval Observatory time transmissions, NAA, it would appear that the assessments of the observatory clocks from both the monitoring station and NRC were somewhat biased by US sources of calibration.

The impact of the quartz crystal frequency standard upon the astronomers was definite. A year after his report that the fourth room in the clock vault was being prepared for a second synchronome, C.C. Smith was turning a friendly ear to the suggestion by Henderson that they assemble the necessary parts for their own crystal clock. Smith retired in December 1937 before any steps could be taken. In fact it was not till the depression years were over and priorities for the war effort were being claimed that a General Radio Frequency standard was obtained for the time service. It arrived at the observatory on 17 April 1942, and became known as Co to distinguish it from other crystal standards, Cm of the Monitoring Station and Cr of NRC.

General Radio had several alternatives to offer. The deluxe model, involving three oscillators, was similar to the one acquired by MacKinnon. An economy unit had only one crystal. The crystal frequency was doubled from 50 kHz to 100 kHz, then divided down in two multivibrator steps from 100 kHz to 10 kHz to 1 kHz. The original crystal purchased by Bain for DOT had a total of four tubes per stage, all of filament type, and power was derived from floating batteries. This caused problems. 'Bias voltages were from "C" batteries which always seemed to be going dead,' Ed Davey later recalled. 'Filament-type tubes were very critical to filament voltage, a drop of two-tenths of a volt often causing one or other of the multivibrators to drop out. Consequently an uninterrupted run of more than a week was a real achievement.' By the time the order was being considered for the observatory, however, a model which could be plugged directly into alternating current was available. Since there was no room for another large bank of batteries, it was decided to take advantage of the modern design of tube rectifier for the plate and grid bias voltages. For economy a single crystal unit was ordered, and its synchronometer provided the seconds pulses by which it was compared with the other clocks of the observatory.

An interruption in alternating current was a real, though not a very frequent, hazard. As a precaution, Henderson installed a dynamotor, which generated 110 volts ac from 32 volts dc. The observatory battery bank was tapped at 38 volts and #12 copper leads were used from the battery room to the time room, a total distance of perhaps thirty feet. High speed (1 rps) chronograph records showed that the build-up time of the generator was about 0.6 second, and the heaters of the tubes seemed to bridge the build up time satisfactorily. The high voltage supply was protected against short interruptions by a large-capacity condenser. In 1952 the motor generator gave way to a Cornel Dubilier vibrator converter of 375 watts capacity, similar to the type used as a source for lighting on railway coaches. Within the next two years a total of five such units were installed, affording individual protection to each of the principal crystal clocks and their associated circuits.

Early in 1948, as a conservation measure, Hydroelectric authorities began cutting power for brief periods in various parts of Toronto; Ottawa received similar treatment later in the year. The observatory, being on the same electrical circuit as the Civic Hospital, was not subjected to deliberate cuts, though adjacent buildings on the Experimental Farm were. However it was disturbing to experience drops in voltage of 10 per cent which seriously affected the output of amplifiers, mainly through reduced cathode emission.

The threat of power cuts, whether deliberate or accidental, was sufficiently real that an emergency gasoline generator was requisitioned. A survey revealed that the minimum power requirements within the time service, including all the clocks, amplifiers, telescope illumination lamps, receivers, relays, and lights for the time room, amounted to 11.55 kilowatts. In November 1948, a Hercules gasoline-powered generator of 20 kilowatts capacity was installed in the basement of the machine shop, and the necessary wiring to the time service was completed early in 1949. About fifteen seconds were required for build-up time, and this was readily bridged by the battery-operated vibrators.

With this improvement, Co (the observatory's crystal standard) became the primary timekeeper of the observatory, with Cm (Monitoring Station) and Cr (NRC) serving as reference standards. s29 was relegated to the lesser duty of providing sidereal time for transit observations. Even in this role s29 was limited, and within a few years it was to be replaced by sidereal seconds pulses derived electromechanically from a crystal source.

Traditionally the time service had relied on battery supply for the operation of clocks, relays, telescope illumination, micrometer contacts, and so on. The battery room, with its banks of large capacity lead acid cells was a source of direct current of any value up to about 100 volts. Every ten years or so the cells had to be replaced. Originally, a motor generator was used to restore the charge once a week. In the early 1930s the generator, now worn out, was replaced by a two-tube tungar rectifier. About this time small copper oxide dry-disc rectifiers were introduced, either to maintain a floating charge to compensate for the drain on a portion of the battery bank, or to provide an independent source of direct current.

As the wiring of the observatory aged, and the insulation deteriorated, leakages to ground tended to develop at different points across the battery. Once it was a wire to the dome for telescope illumination. Then it was a wire to the transit annex for the operation of a relay. A third case was a wire which provided field illumination for the transit telescope; and there were more. The requirements for six volts here, eight volts there, throughout the observatory were distributed uniformly across the battery bank, and a short to ground could occur with disturbing effects almost anywhere. Soon after the second world war the old wiring, consisting of bundles of twisted pair lamp cord which spilled out of overcrowded wooden troughs and festoooned the hallways, was replaced by metal-clad multi-wire cables. Terminal blocks, properly identified, greatly simplified the task of co-ordinating and monitoring all the services from the time room throughout the observatory and the city.

Returning to the General Radio frequency standard: other problems besides voltage supply required attention. The crystal oscillator would not hold its frequency in spite of attempts to adjust the inductance and capacity of the associated tuned circuit. The crystal had to be replaced, and in December 1944 a new 50 kHz quartz bar was installed. One instance will suffice to show how several troublesome factors will converge. In 1948 a power failure of 1 1/2 minutes was not bridged, apparently because the battery bank had been allowed to run down. But quite independently several tubes had run their normal course and had to be replaced before Co was operational again. The philosophical outlook on these various mishaps was that experience was gained on how to avoid a subsequent occurrence. More important, though, was the fact that a single frequency standard was quite insufficient for the effective operation of a time service. The other two standards, Cm at the Monitoring Station and Cr at NRC, were very useful in assessing the performance of Co, but neither of them could be readily switched into service in place of Co. More crystal standards were required at the observatory.

Just a year after this episode with Co, s29 came almost to a standstill when the impulse from the slave failed to arrive because of a blown fuse on one of the several outputs of the battery bank. The slave pendulum was easily restored, but the free pendulum was more difficult. When all but one of the mounting bolts were removed in order to tilt the case the vacuum seal developed a leak. A week elapsed before the seal was restored. Meanwhile Time Machine No. 1 was switched to Co control, and it would have remained that way indefinitely were it not for the faulty relay which two months later caused the output to slip several seconds; TM1 was thereupon restored to s29 control.

In 1947 Hollinsworth of the Time Service had completed construction of a multivibrator chain which counted down from 1000 hertz to 100 hertz, from 100 hertz to 10 hertz, and then, with a multiplication of six, generated 60 hertz. Much of the pioneer work had been accomplished by Henderson a decade earlier, but had never been engineered into a convenient and compact rack-and-panel assembly. Hollinsworth had some difficulty with the 10 hertz unit because of the bulky components and because of division by ten at the low frequency. Working on another angle, McKinley achieved a crystal control of 60 hertz by using a six-sector disc of a nickel-iron alloy called mumetal as an inductor on the 10 rps shaft of the synchronometer.[8] On account of occasional stoppages the synchronometer was not as consistent as the multivibrator chain, but either source could be fed into the 60 hertz amplifier which produced 150 watts at 110 volts, 'sufficient to operate quite a number of small motors drawing from 2 to 10 watts.'[9]

Hollinsworth then designed a sidereal converter employing two gear ratios, 119:114, and 317:333. The input was a 1 rps crystal-controlled 60-hertz motor. The 1 rps sidereal output differed from ideal by about one second in eight years. There was both a phasable contact, whose position could be read to the millisecond, and a fixed-phase output contact, so that the converter could be used to compare the time of other sidereal clocks and also be a sidereal clock itself. An attempt was made to derive a 60-hertz sidereal by using a 60-sector mumetal disc on the 1 rps shaft, following the example of McKinley. But the frequency modulation resulting from the residual eccentricity of the gear wheels made the result impractical.

Henderson used the principle of a rotary transformer to derive 60 hertz sidereal. The 60-hertz mean time was split into three phases, using a

8　Bourne and McKinley, 'A pulsed crystal oscillator range calibrator' and 'A precision radio time signal system'

9　V.E. Hollinsworth, 'On the use of crystal controlled synchronous motors' condenser advance, an inductive retard, and a resistance for zero phase, and the three wires were fed into the field windings of a selsyn motor. The output from the rotor winding could then be advanced or retarded according to the direction of rotation. With a simple gear ratio of 23:140 on the output of a one rps mean time shaft, the selsyn shaft was advanced 0.16428 of a revolution each second. The resulting 60 hertz had a gaining rate of 0.016 second per day on sidereal time. Not only was it a small rate, but it was far more uniform than could be expected from the best pendulum. As a result, electric clocks driven by this frequency replaced the old sidereal seconds dials that had been used as observing monitors in the transit rooms and the domes, and electric motors replaced gravity drive for chronographs and equatorial telescopes. In reporting the 1950–3 observing program with the meridian circle transit telescope, R.W. Tanner and E.G. Woolsey stated: 'The printing chronograph has been converted to synchronous motor drive on crystal controlled frequency, and served as observing clock with small nearly constant rate. Hourly comparisons of the chronograph were made, at first with the slave of Shortt 29, later with a mechanical converter controlled by a second mean time crystal.'[10]

Haines Scientific Instruments of Englewood NJ produced an electric clock with an internal gear ratio from mean time to sidereal. Several of these were installed as monitors operating from wall outlets in places where sidereal time was required with precision to the minute only.

C.S. Beals, who had become Dominion astronomer in 1946, was careful to consult with McClenahan and members of the time service staff in planning the future. In spite of moves at Washington and Greenwich observatories to rely upon crystal frequency standards, McClenahan felt some reluctance at abandoning the pendulum. It had the advantages of long periods of trouble-free operation, low power consumption, and precision adequate for most timekeeping requirements. Younger staff members, however, carried the day, and in the first years of the 1950s additional crystal primary frequency standards were acquired.

The first, from General Radio, arrived in April 1950. An updated model of the one bought in 1942, it had a 100 kHz crystal and improved thermal control of the oven; a synchronometer provided an output at the seconds level. The next two were Western Electric frequency standards, and arrived in 1951 and 1953. McKinley of NRC already had one of these

10 E.G. Woolsey and R.W. Tanner, 'Results of observations made with the reversible meridian circle 1950–7'

and had been impressed by the calibration of the oscillator circuit which permitted an adjustment to 1 part in 10^9, equivalent to about a tenth of a millisecond a day. The Western Electric standards were complemented by Ernst Norrman multivibrators, which counted down from the 100 kHz frequency of the oscillator and provided outputs at 1000 hertz, 100 hertz, and 60 hertz. A General Radio synchronometer for each standard provided a realization of the time and day-by-day comparisons at the seconds level. Ernst Norrman arrived at 60 hertz by counting down to 100 hertz, multiplying to 300, and dividing down to 60, thereby avoiding the problems Hollinsworth encountered at 10 hertz. Because of the relative ease of adjustment, the Western Electric units became the primary time standards and were removed to the clock vault for better control. With four crystals, the old name 'Co' ceased to exist. Instead, the crystals became known as x1, x2, x3, and x4, the first two referring to the two General Radios and the others being the Western Electrics.

Various techniques were employed to compare the performance of the crystal clocks. Henderson devised a scheme in which the minute pulse operated a light metal pointer like a hammer onto a drum chronograph. The worm which advanced the carriage was reduced in speed so that the record could be accumulated for four days. At first a carbon paper was wrapped around the drum on top of the chronograph paper. But it was just about the time that ballpoint pens were introduced, and a ballpoint cartridge was used to mark directly onto the paper. The resulting chart showed the short-period as well as the long-period variations of each of the crystals. The method was not continued after Henderson reached retirement age in March 1956; continuous strip chart recorders, designed to give similar information, were coming into service, and in the next few years several were acquired.

In the early 1940s a ring crystal frequency standard had been developed by L. Essen and D.W. Dye in the Research Laboratories of the British Post Office. A decade later several models were ordered by different sections of NRC. In the Dominion Observatory McClenahan, Beals, and J.P. Henderson were sufficiently impressed that an order for one was placed. It was to operate like an astronomical pendulum with no adjustment in rate once it was factory tested, and the frequency trimmer was accordingly left out. All the components except crystals arrived early in 1954, and in July of that year Jack McClements of the British Post Office brought the crystals by hand and assisted in their installation. The observatory ring became known as R5 or 5DO.

One day McClements and Lloyd Miller were carefully observing the beats between the new ring crystal and the General Radio crystal x2 which was located in the time room. They were seated quietly in the wireless room next door. Suddenly one of the summer students bounded across the time room floor because her boyfriend had arrived unexpectedly, and with each step the frequency of x2 shifted noticeably. So the calibration of the Essen ring was held up until x3 (a Western Electric in the vault) was substituted. But the little incident actually demonstrated that the General Radio crystal was not sufficiently shielded from vibration. NRC came to the rescue with a shock mount, and both x1 and x2 were removed to the vault [Plate 51]. The Essen ring was also placed in the clock vault, the very room which C.C. Smith, some twenty years earlier, had designated as the home of a second Shortt clock. Coaxial cables from the clock vault to the time room connected the various oscillators to the appropriate multivibrators, amplifiers, and other gear.

The Essen ring multivibrator counted down to 50 hertz from which a small 50-hertz motor was driven. A single tooth on the 1 rps shaft advanced the second hand of the clock in one-second jumps, resulting in an audible click. Because of the noise it created, the clock was not used. Furthermore the rather high negative rate of the oscillator really made it of little value as a clock. The rate continued to increase negatively, and never approached zero, as had been forecast.

McClements gave Hollinsworth and Miller the circuit details for a multivibrator using cold cathode tubes to count down from 50 hertz to one pulse per second. Several units were made up and adapted to the Ernst Norrman and General Radio low frequency outputs. The resulting pulse was more uniform than the synchronometer contact and provided an effective way of measuring the accumulated daily gain or loss of each crystal with respect to the others to a tenth of a millisecond a day. Already this amounted to a hundredfold increase in precision over the pendulum clock.

The British Post Office had developed a circuit which used the ordinary watt meter to count the beats between two oscillators. They were made up in units of three to display the continuous performance of three crystals at the 100 kHz level. Two of these units were installed in the observatory time room in 1956, and two groups of three crystals were made up by using the ring crystal R5 twice. A polaroid camera was mounted across the room to record photographically the reading of the dials at 10:00 A.M. each day. The relative performance of each crystal could then be read to a very high precision, but the short-term variations did not warrant using the value beyond 1:10.[10]

More of the ring crystals were acquired by the Ottawa group within the next year or two, bringing the total to nine. There were four at the Applied Physics Division of NRC, three at the Radio and Electrical Engineering Division of NRC, and two at the observatory. More than ever it became evident that this resource could be pooled to provide a Canadian standard of time and frequency.

Part of the ground work for the intercomparison of these crystal standards had been laid more than a decade earlier when in 1944 a private telephone line had been installed between the observatory and the home of J.P. Henderson in order that the radio receivers could be removed from the electrically noisy environment of the observatory. The receivers remained switched on, each tuned to one of the time signal stations, and available by a selector switch remotely controlled from the wireless room at the observatory.

The line provided a further avenue for research and experimentation. When 60 hertz was available from the frequency standard, Henderson superimposed it on this line to find out if it could be detected and used without being contaminated by the 60 hertz of the power lines. The experiment was a success. He ran a clock at his home for several weeks from this source, and a clock was then operated for a short while in the CBC studio in the Chateau Laurier. But the system did not survive because there were too many uncertainties in the electronics as well as in the telephone line itself.

In 1953 the telephone line was used to test successfully the transmission of 100 kHz from the observatory to Henderson's home, following which this frequency was sent by telephone line to NRC on Sussex Street to intercompare the ring crystal timekeepers of the observatory with those at the Council. It is not known just how 100 kHz travelled through the Bell Telephone system. Theoretically the frequency was too high for that type of transmission. It was also too high to cause any interference to the normal telephone circuits, and therefore could be imposed onto the line at ten volts, which was higher than normal. It arrived at NRC Sussex Street as a weak signal of 500 microvolts and was detected by a radio receiver. C.F. Pattenson used the same method to provide monitoring in the Applied Physics laboratory of the performance of his Essen ring crystal. A 10 kHz line from the observatory to Pattenson's lab on the Montreal Road provided a satisfactory way to close the loop between the three labs. Bell Telephone assisted in this network by removing all repeaters and ensuring that continuous copper wire was on each line.

The observatory's ring crystal standard R5 was assessed with respect to mean solar time, UT2 (see appendix), from observations with the photographic zenith tube, and the results communicated to the other two labs. S.N. Kalra, of the Applied Physics Division of NRC, then undertook to establish a Canadian standard of frequency based on the evidence of the frequency standards located at the three centres. It was realized, however, that the accuracy of the absolute frequency determination was limited by the errors of astronomical observations and the uncertainty of the corrections to be applied. In most cases it was considered possible to achieve an accuracy of $5:10^9$ if it were assumed that all the corrections needed for the determination of UT2 were known exactly. In addition it was known that there were secular changes in UT2. As if to compound Kalra's problems, the oscillators were found to be subject to unexplained frequency shifts of 1 part in 10^9, thus limiting the precision with which they could be used to interpolate and extrapolate time and frequency.

A contribution to the crystal oscillator era of timekeeping in Ottawa came from an unexpected quarter. H.D. Valliant of the Dominion Observatory Gravity Division had spent a frustrating field season because the portable crystal oscillator and associated multivibrator stages which he was using had been difficult to maintain. In preparation for future field work he acquired in 1959 a James Knights 1 MHz crystal frequency standard, fully transistorized and complete with a standby floating battery supply. This he took to Kalra for an assessment, where it measured up to all its advertised specifications. From Kalra it came back to the observatory and was used for several months as a part of the time service. So impressed were Hollinsworth and Miller that one of the instruments was immediately requisitioned and became the controlling oscillator for CHU. Subsequently the older crystals at the observatory were retired as more efficient and more precise oscillators of the James Knights–Sulzer design were acquired.

ATOMIC TIMEKEEPERS

It had been hoped that the Essen ring crystal oscillator would have served as a very precise frequency standard, particularly when a group of five was initially installed in the Ottawa laboratories, the number later being increased to nine. The history of their performance in Great Britain at the National Physical Laboratory (NPL) during and after the second world war had been good, and at the time of purchase they were the finest on the market. However, as mentioned earlier, an attempt to use them as a time base to detect variations in earth rotation was not successful.

Even as the ring crystal was being pushed to the limits of its capabilities, advances were recorded in atomic physics which had a profound effect on timekeeping. As early as the 1930s, the idea of a perfect clock, based on a natural frequency of the atom, had appealed to the imagination. Advances in atomic physics during the war paved the way. In the 1940s a clock was built using an absorption line of ammonia. Other experimenters devised a method for successfully utilizing the resonance due to the hyperfine splitting of the Cesium 133 atom.

The first device based on this principle to be used as a frequency standard was placed in operation by L. Essen and J.V.L. Parry at the NPL at Teddington, England, in 1955. It gave the value of the mean solar second as 9,193,631,830 ± 10 cycles of the frequency of the central absorption line. In a three-year co-operative program between NPL and the United States Naval Observatory, Markowitz, Hall, Essen, and Parry determined the value of the ephemeris second as 9,192,631,770 ± 20 cycles of the frequency of the central absorption line of the cesium 133 atom.

The National Company of Malden, Massachusetts, produced in 1956 what they called an Atomichron, a crystal oscillator whose frequency was controlled by a cesium beam tube. It was the first of a number of Atomichrons built within the next few years. Two units were taken to NPL and placed alongside the laboratory standard built by Essen and Parry to study the differences which caused a small discrepancy in output frequency.

Canadian scientists, aware of the progress, were quick to develop a research program. Work on a cesium resonator similar to that of Essen and Parry was undertaken at NRC by S.N. Kalra, R. Bailey, and H. Daams in 1956, and the resonator was brought into operation in 1958. The accuracy of the first NRC cesium standard, CsI, was several parts in 10^{10}. It immediately became the Canadian standard of frequency to which the observatory clocks were referred daily.

When Kalra left NRC, the team leader became A.G. Mungall, under whose leadership CsIII was developed and set in operation in 1965 [Plates 52 and 53]. Initially it had an accuracy of a part in 10^{11}, but this was subsequently improved to a part in 10^{12}. CsIV was an experimental model incorporating ideas which were not successful. Currently under test is CsV, developed by the team of A.G. Mungall, R. Bailey, H. Daams, D. Morris, and C.C. Costain [Plate 54]. A greatly improved resonator built on the same general design of CsIII, it indicates an accuracy of one or two parts in 10^{13}. In addition it is designed to operate continuously as a clock, and not be restricted to the role of a primary frequency standard. With this in mind it is equipped with a large capacity 24-volt battery, float-operated, with 24-volt inverter power supplies, all augmented by diesel generator standby.

In 1967, after some four years of research at NRC, two hydrogen masers were brought into operation [Plate 56]. This class of atomic oscillator, because of its excellent short-term frequency stability of better than a part in 10^{14}, is an invaluable tool for assessing the performance of the cesium resonator. It is hoped to extend the period of maximum stability, which at present is about one thousand seconds, and perhaps discover a way to use the hydrogen maser as a primary frequency standard.

From about 1965 the atomic time scale at NRC was supported by a group of four 2.5 MHz crystal oscillators. The best of these was calibrated daily with respect to CsIII, and all were continuously intercompared. Digital divider chains from the crystal oscillators produced seconds pulses for the physical realization of time scales. In 1968 a Hewlett-Packard 5061A cesium standard was acquired. This, measured at weekly intervals with respect to CsIII, proved to be greatly superior to a crystal oscillator as an instrument for interpreting the value of the cesium resonator and displaying atomic time.

Meanwhile at the observatory a rubidium vapour frequency standard by Varian Associates was acquired in 1963. This was followed in 1966 and again in 1967 by the acquisition of two Hewlett-Packard cesium standards, and in 1969 with a third. The rubidium standard was moved out to the transmitter laboratory in 1966, but was transferred to standby duty in 1968 with the installation of a Hewlett-Packard cesium standard as the time and frequency control of CHU, the radio transmitter.

When the control of CHU carrier frequencies and seconds pulses was synthesized from a 1 MHz crystal oscillator by J.C. Swail and C.F. Patterson in 1959, provision was made to monitor it by direct wire so as to maintain the oscillator within a part in 10^8. The success of the operation permitted Ottawa to join with other national time services in a program launched at the end of 1960 by Greenwich and Washington whereby the seconds pulses transmitted by each institution were synchronized to the nearest millisecond. Time co-ordinated in this way was called UTC (Coordinated Universal Time). The next step in precision control of CHU was made with the acquisition of a James Knights 2.5 MHz transistorized crystal frequency standard with a battery-floated power supply. One was installed at the transmitter lab in 1961, providing a stability of a few parts in 10^{10}. After five years of good service it became standby to the rubidium atomic standard.

By the time the third Hewlett-Packard cesium standard had arrived at the observatory in June 1969, conversations had already commenced concerning the consolidation of government responsibility for astronomy under the National Research Council. It was as a result of these discussions, and also as a fitting climax to the several years of close collaboration between the time service group of the observatory and the frequency group at NRC, that they were brought together to form the Time and Frequency Section of the Physics Division of NRC on 1 April 1970. Paradoxically, the only part of the time service that did not move across to NRC was that part, headed by R.W. Tanner, which was most astronomically oriented, namely the PZT group. In the Canadian context the PZT observations of time and latitude, which revealed a variable rate of earth rotation and wander of the pole, had within the previous year or two contributed more to geodynamics than to timekeeping. Time was now the responsibility of the physicist, and the second was defined in terms of the atomic standard, of which NRC was henceforth the Canadian custodian.

The actual operation of the Canadian PZT was unaffected. Observations continued to be made every clear night at Ottawa and Calgary, and the results of the plate measurements, which yielded time and latitude, continued to be forwarded on a weekly basis to the Bureau International de l'Heure and the International Polar Motion Service. Also, close liaison was maintained with the Royal Greenwich Observatory at Herstmonceux because of the twinning of the Calgary PZT, it having been established on the same latitude as the older instrument in the United Kingdom. In Ottawa, reception of the ground wave CHU transmissions continued to assure a close attachment of the PZT observations to the NRC atomic time scale. The PZT group remained with the Earth Physics Branch, one of whose interests was the study of motions in the crust of the earth.

The introduction of commercially designed atomic timekeepers brought in its wake the phenomenon of the flying clock. The older method of comparing time at a distance by the monitoring of radio signals remained valid for the comparison of time interval. For the comparison of time itself as opposed to time interval, the radio link was inadequate.[11] There was a loss of precision and purity of the radio wave, and there was an uncertainty in the travel time of the radio signal. It had been tolerable when uniformity to the millisecond was satisfactory, but now, with microsecond precision within each lab, the only satisfactory method

11 It has always been possible to measure with precision the interval between two pulses that originated at a distant station, but without some further evidence it has not been possible to say with precision how long it took for the pulses to travel the distance.

of intercomparison was to transport a timepiece with the same characteristics of accuracy and precision.

One of the first flying clock experiments was reported by the US Army Research and Development Laboratory at Fort Monmouth NJ. In November 1959 an Atomichron was test flown in an aircraft, and the following February it was used to compare Atomichrons in Fort Monmouth and Patrick Air Force Base, Florida. The experiment demonstrated the feasibility of comparing atomic clocks to the microsecond on a world wide basis. In April 1963 two scientists from the National Bureau of Standards transported a high precision quartz crystal clock between Boulder, Colorado, and Greenbelt, Maryland, establishing a synchronization of clocks within ±5 microseconds.

Hewlett-Packard of Palo Alto, California, absorbed the National Company of Malden, Mass., and became the major producer of cesium clocks during the 1960s. In 1964 two of their cesium clocks were made portable and flown to Europe to compare the time between Neuchâtel, Switzerland, and Boulder, Colorado. The time difference was established to about a microsecond, and the travel time of the very low frequency (VLF) propagation between Boulder and Neuchâtel was evaluated with an uncertainty of about 200 microseconds. Also, the frequency of the long-beam cesium standard at Boulder was compared to the one at Neuchâtel within a few parts in 10^{12}. Similar experiments were conducted by Hewlett-Packard for three successive years, the distance travelled being extended and the instrumental equipment improved each time. In 1967 the experiment required forty-one days, during which fifty-three places in eighteen countries were visited.

The Dominion Observatory and NRC were included in these flying clock measurements in 1966 and 1967, thereby relating the time and frequency at both these laboratories to the visiting flying clock, with an accuracy of a tenth of a microsecond for time and a part in 10^{12} for frequency. Early in the summer of 1967 the Swiss company Ebauches transported a cesium standard from Switzerland to Montreal to serve as the control timepiece for Expo 67; on the way the clock was brought to the time lab of the Dominion Observatory for a time and frequency comparison [Plate 55].

In 1960, Atomichrons had been placed as frequency controls at the Omega navigation stations at Oahu, Hawaii, and Forestport NY to improve air and sea navigation. Subsequently atomic standards have replaced quartz oscillators at all Omega and Loran-C stations. The network of Loran-C stations, maintained as navigational aids by the US Navy, now involves the use of large numbers of cesium clocks, which require monitoring by regular visists with a portable clock. In the performance of this duty, the US Naval Observatory field team commenced in 1968 to pay regular visits to Ottawa. Not only did these visits establish a time comparison to the tenth of a microsecond between Washington and Ottawa labs but they also established with the same precision the travel time of the Loran-c signal. Special receivers designed to identify the pulses of the transmitted signal were then used to compare both the time and frequency of the laboratory standard with the Loran-c station. These observations, when reported to the Bureau International de l'Heure, have provided a means for establishing a world clock and intercomparing all the laboratories who reported similar observations.

The feasibility of time comparison at a distance by satellite link had been demonstrated in two separate experiments, one in 1964 across the Atlantic from the USA to England, and the other in 1966 across the Pacific from the USA to Japan. The method has not been pursued, though it may have value for the future.

The unit of time, the second, has been defined as a specific frequency corresponding to the transition between the two hyperfine levels of the ground state of the cesium 133 atom. Seven laboratories contributed to the initial establishment at the BIH of a uniform frequency and time scale. Three of them, the National Research Council of Canada, the National Bureau of Standards of the USA, and the Physikalisch-Technischen Bundesanstalt of Germany, have long-beam laboratory standards. There are now twenty-three collaborating laboratories. Since, except for these three labs, reliance is placed upon the evidence of commercial cesium standards, most of them built by one supplier, Hewlett-Packard, there is the possibility of a common bias and thus of a systematic drift in the internationally adopted atomic time. NRC, with its own CsIII operating and CsV soon to become operational, would have two independent standards operating and thus be in a position to make a unique contribution to the realization of the frequency of the cesium transition and hence to the evaluation of the atomic second.

Two further methods of clock synchronization have recently been developed and will be mentioned briefly. They are the aircraft overflight experiment, and the TV synchronization link. Each was developed to serve a local national need, but the first one has global significance. The Office National d'Etudes et des Recherches Aérospatiales (ONERA) of France had developed the technique for synchronizing the clocks at several observing sites in the Mediterranean area in connection with rocket launching. An aircraft equipped with a cesium standard and microwave equipment for time pulses would fly over each site at a precise altitude. With matching equipment on the ground, the several sites would have their local standards related to each other within about ten nanoseconds. In September 1970 ONERA undertook to conduct the more extensive experiment between Europe and North America. Matching ground equipment was installed in advance and the aircraft flew in succession over the time laboratories of the Observatoire de Paris, the Royal Greenwich Observatory, and NRC. Security regulations prevented a direct flight over the US Naval Observatory, but a two-way exchange from a nearby airport accomplished the same result. The four time scales were compared to a few tens of nanoseconds, an improvement of about one order of accuracy over that currently available with the portable clocks.

The television synchronizing pulse was identified as a very precise time signal by laboratories and industries within range of the same TV program. First developed in Europe, it was recognized by the National Bureau of Standards as a vehicle whereby time across the USA could be synchronized. The three major American TV networks are each controlled by a separate rubidium atomic frequency standard, and their nation-wide programs are programmed through their New York studios. The synchronizing pulses are monitored in the USA at NBS in Boulder and USNO in Washington, and in Canada at Ottawa, with results that are comparable to Loran-C. Ottawa is close enough to the US border to be able to monitor one of the American TV outlets.

A similar plan was not practicable in Canada because there was no central studio from which TV programs were broadcast nation-wide on a regular basis. The 4½-hour time difference from Newfoundland to the Pacific coast is just too wide a spread, except for some special event such as the annual Christmas message from the Queen. However, the CHU transmitting station is well within range of an Ottawa TV signal which can also be received at NRC. As a result the CHU time signal is now maintained within a microsecond of the NRC time scale, synchronization having been established on one of the visits of the portable cesium standard of the US Naval Observatory.

The NRC primary cesium standard, CsV, operating since May 1975, had by January 1977 exceeded expectations. With an accuracy of 5 parts in 10^{14}, it had replaced CsIII as the Canadian standard of frequency. Furthermore, because of its novel design it was operating continuously as the world's first primary laboratory-type cesium clock. Quite justly it might be called the most precise timepiece in the world. During 1976 it accumulated an error of less than half a microsecond compared to IAT (International Atomic Time). It has set a pattern which will likely be followed by other standards laboratories.

Recently NRC, along with the only two other laboratories maintaining primary laboratory cesium standards, the National Bureau of Standards in the United States and the Physikalisch-Technischen Bundesanstalt in Germany, agreed that the value of the cesium second needed to be adjusted by a part in 10^{12}. This was done at the beginning of 1977, and would result in a change in international atomic time of a thousandth of a second in about twenty-eight years! The outstanding success of CsV inspired the hope that CsVI, a compact working cesium standard based on the same design, would have a similar high precision. Three units were to come under test late in 1977.

6

Radio time signals

When, from the Dominion Observatory itself, we look outward to consider its role as a timekeeper for Canada and the world, the contribution made by the rise of radio time signals is seen to be of central importance. Of this revolution in communications the aspect we shall examine first is the method developed at the observatory to monitor and use radio signals as time checks. We shall then go on to review the observatory's own use of radio transmissions to distribute time across Canada.

The Dominion Observatory had been one of the first institutions to achieve success in determining longitude by using wireless time signals. In April 1914, W.F. King, the chief astronomer, had advocated, as a survey experiment, the erection of a temporary receiving station near Haileybury in northern Ontario 'as nearly as possible under field conditions.' In a letter to the Radio Branch of the Department of Marine and Fisheries he wrote:

A chronometer has been installed in the wireless station at Kingston, which can be used for sending out signals whenever required; it is proposed in the preliminary experiments to determine whether these signals (or, failing them the Arlington longitude and time signals) can be successfully received with the required accuracy at as great a distance as Haileybury.

Our present proposal is to erect a temporary aerial of say 500 feet to 750 feet length, composed of two aluminum wires supported by portable 30 feet masts lashed to trees as high as possible. I should be very glad of any suggestions ... in regard to ... earth net, guy wires, etc. ... It is desirable to limit the equipment as far as possible to such as could without undue difficulty be transported by canoe, as some of the stations ultimately required can, I believe, be reached only in this way.[1]

The site selected for the first test was Quinze Dam in the headwaters of the Ottawa River. In order to avoid the roar of the rapids, the wireless set was placed in a settler's root cellar. It was a crystal set, built before the days when radio tubes were a common commodity. A similar set was installed at the observatory. Unfortunately the signals from the Barriefield station at Kingston could not be heard at Quinze Dam in spite of all these precautions, and it was necessary to use the stronger signals from Arlington, Virginia. Nevertheless, the success of the experiment that summer demonstrated that wireless time signals could yield results comparable to the telegraph method. The surveyor was no longer restricted to the telegraph line. His instruments could be carried into the vast Canadian northland, anywhere, just so long as he was within range of a radio time signal.

The summer of 1914 was not like any other. Each evening in the field the radio operator, Wally Dier, a skilled telegrapher, copied the news from Sayville, New York, keeping the camp informed of the details leading up to the outbreak of war, then of the progress of the conflict. Some items he copied were messages in English sent to Germany and to German ships at sea warning them of circumstances of the war. When the words came out in German he kept on copying, though he did not understand a word. When the copy was relayed through R.M. Stewart to the Defence Department these special messages ceased, and Wally was officially thanked for his contribution to the war effort.

Four stations in the upper basin of the Ottawa river were occupied during that summer. The following season seven in Quebec and two in Ontario near Georgian Bay were similarly set up. But in 1916 only one position was determined by radio time signals, following which the exigencies of war drew all such activities to a temporary halt.[2]

The end of the war brought a renewal of scientific activity. The Dominion Astrophysical Observatory in Victoria BC, with J.S. Plaskett as direc-

1 W.F. King, chief astronomer, to C.P. Edwards, Naval Service, Department of Marine and Fisheries, April 1914, Department of the Interior correspondence file, Public Archives of Canada
2 Background information is from Klotz Diary.

tor, was officially opened. The Geodetic Survey, which had been a part of the observatory until King's death in 1916, now functioned as an independent unit with Noel Ogilvy as director. Several men who had formerly been attached to the radial velocity program of the 15-inch equatorial telescope were now assigned other duties, requiring the acquisition of new staff.

In July 1919 J.P. Henderson was appointed to the observatory to assist with the radial velocity program.[3] Henderson soon realized that the 15-inch telescope had just about reached the end of its usefulness as an instrument for determining radial velocities. The brighter stars to which it was restricted had all been observed, and fainter stars required very long exposures with such a small aperture. The work was done much more effectively by the new and larger 72-inch telescope in Victoria. Furthermore, his attention had been caught by the attractive program of wireless time signals upon which R.M. Stewart had recently embarked. Henderson had long been interested in radio. At the time Wally Dier was struggling with a crystal receiver at Quinze Dam in 1914, Henderson, a graduate student at the University of Toronto, was spending the summer as a wireless operator on a ship on the Great Lakes, mainly to become proficient in the use of Morse code. He was an experimenter, having acquired some of the early radio tubes in the days when the crystal detector was the principal element of a receiver.

Without neglecting his obligation to astronomy Henderson found himself drawn into the wireless program. Stewart, who had great skill as a mathematician, and some as a mechanic, was quick to recognize the interest and ability that Henderson possessed in the new science of radio engineering. Within five months he had entrusted him with a key to the wireless room, and in return was delighted to see his old tube set augmented by a much more efficient tube set which brought signals in from Washington (NSS), Canal Zone (NBA), and San Francisco (NPG). From then on there was no turning back. In his spare time during 1920, Henderson spent many hours in the wireless room improving the receiving apparatus and devising a method for recording the signals on a drum chronograph [Plate 49]. Also he became associated in a volunteer capacity with C.P. Edwards, Colonel A.W. Steele, and others conducting radio research in the Department of Marine and Fisheries.[4]

3 Klotz records the appointment of Henroteau and Henderson to the staff of the Observatory in 1919.
4 J.P. Henderson diary.

The following year the observatory was called on to play an important role in the survey of northwestern Canada. Oil had been discovered about sixty miles north of Norman on the Mackenzie River, and an immediate survey was needed to locate the numerous claims that were being staked. What this implied was described by A.H. Miller, a member of the observatory's field party:

It would have been too great and too expensive an undertaking to extend so far north in such a short time the meridians and base lines which in Western Canada form the control for subdivision surveys and traverses. On the other hand it was most desirable that the surveys in the far north should be linked up with the existing surveys in Western Canada. Consequently it was decided to run a stadia traverse from the most convenient northerly point in the existing surveys right down to the oil fields. The part of the Dominion Lands survey that was chosen from which to begin the traverse was the point where the 30th base line crosses Slave River. This is about 65 miles up the river from Fitzgerald at the head of Smith rapids and the approximate latitude of the point is 59° 08′. From here to the oil well below Norman the distance is nearly one thousand miles. In running such a long traverse it is only to be expected, in spite of the most careful work, that small errors will creep in and that their effect may accumulate as the survey progresses. In order to correct these errors and to prevent their accumulation certain suitable points in the course of the traverse were selected at which the latitude and longitude were to be determined. The traverse was 'tied in' to these points as it proceeded and the position of all points along the survey are therefore referred to the geographical positions of the selected points.[5]

The observatory's job was to find the latitude and longitude of each of these control points. The first part was accomplished during the season of 1921, when the survey was carried down the river about one hundred miles below Norman. The next summer positions were established as far down as Arctic Red River as a basis for the survey of the Mackenzie Delta. With no telegraph stations available, radio time signals were used.

The observatory party was made up of three persons, H.S. Swinburn, who took the star observations, J.P. Henderson, who secured the wireless time comparisons, and Miller, who made measurements of the force of gravity at each of the stations.

Swinburn was supplied with a two-inch astronomical transit of the broken type, fitted with a transit micrometer, and also a sidereal

5 A.H. Miller, 'Dominion Observatory work in the Mackenzie District'

chronometer and a drum chronograph. His work was hampered somewhat by the short period of darkness in the arctic summer nights and by
the mists that frequently rose from the land of lakes and muskeg. It thus
required from ten days to three weeks to complete the observations at
each station – after the cement piers had been constructed. For his gravity
measurements, Miller used a half-seconds (invariable) pendulum apparatus which had been in the possession of the observatory since 1902.

The use of radio signals was outlined in Miller's report:

The receiving of the wireless signals was attended to by Mr J.P. Henderson. He
also had difficulties to contend with. Owing to the great distances from the
sending stations and the almost continuous daylight, static disturbances gave a
considerable amount of trouble. It was therefore necessary to set up the wireless
apparatus to the very best advantage. Although at points farther down the river it
was not possible to get very much of the press news sent out by wireless, Mr
Henderson was very successful in getting the time signals from the various
stations, which were absolutely necessary for the successful completion of the
work. At the times the signals were received they were compared by the method of
coincidences by extinction with a chronometer which was set to gain about one
second in every hundred seconds on the mean time signals. As the signals last for
five minutes there are ordinarily at least three times during the five minute
interval when the second beats of the signals coincide with the beats of the gaining
chronometer. It is possible by this coincidence method to make a time comparison
to within a hundredth of a second. This chronometer was compared on the
chronograph, each time just before and just after the receipt of the signals, with
the sidereal chronometer used by Mr Swinburn in his star observations. In this
way the sidereal chronometer time of arrival of the signals at any particular station
was determined to within perhaps a few hundredths of a second. As the sidereal
chronometer was in turn compared (also on the chronograph when the star
observations were made) with the time of transit of the stars, it was finally possible
to determine the arrival of the wireless signal in terms of the sidereal or local star
time at the place of observation. The same signals that were received at the field
stations were also received at the observatory at Ottawa and their time of arrival
was compared with the standard Riefler clocks of the observatory. The variations
of these clocks from the local sidereal time at Ottawa are determined very accurately from the regular observations that are taken with the meridian circle. The
local sidereal times of the same time signals were therefore determined both at
Ottawa and at the field station. After allowing for the time of transmission of the
wireless signal (about one hundredth of a second for 2000 miles) the difference

between these local times for the arrival of the same signal gives the difference in the longitude between Ottawa and the field station.

It was a heavy daily schedule that Henderson set for himself: 1 A.M. NPG (San Francisco), 5 A.M. NBA (Darien, Canal Zone), 12 noon NSS (Washington), 1 P.M. NBA (Darien, Canal Zone), 3 P.M. NPL (San Diego), 10 P.M. NSS (Washington). The times were Eastern Standard, for the convenience of comparison with results of signals observed at Ottawa. In addition to the routine of wireless time signals there was the maintenance of the radio equipment, batteries, receivers, and antennas. The news, so vital to avoid isolation, was copied whenever possible, sometimes competing against the static which more often than not determined the end of reception.

In 1921 five stations were occupied, Peace River, Providence, Simpson, Norman, and Resolution, and the following season four more, Liard, Good Hope, Arctic Red River, and Chipewyan. The operation was a success from every standpoint, a fulfilment of the forecast made in 1914 that radio would free the surveyor from the limitations imposed by the telegraphic method of time exchange. In addition, by copying the news broadcasts at every opportunity, Henderson demonstrated that radio could bridge the gap and break down the isolation of the communities of the north. An attempt had been made to extend the telegraph, but only a relatively short distance had been covered by 1922, and the cost of servicing the poles, which could not be held upright for long in muskeg, was high. The 1921 and 1922 field excursions by the observatory survey team demonstrated the effectiveness and efficiency of radio communication, and shortly thereafter Col. A.W. Steele authorized the establishment of a radio communication network for this northern frontier.

The survey party had enjoyed giving extra-curricular demonstrations of the power of radio to northerners. One day in 1921, amid heavy static, Henderson recorded the result of the boxing match in which Georges Carpentier was defeated by Jack Dempsey. As Otto Klotz later reported: 'At Fort Providence, in the wilds of the Mackenzie Basin, were three Dominion Observatory astronomers recently observing for geographic positions in connection with oil discoveries ... The wireless expert and astronomer, J.P. Henderson, caught the wireless message announcing the result of the encounter between Dempsey and Carpentier, showing the latter to be an eclipsing variable, radial velocity in line of sight 120 km a second, period 12 minutes for 4 revolutions. The news was had at Fort

Providence as soon as elsewhere, and spread up and down the Mackenzie to the consternation of natives and others, as this is the only wireless outfit in these boreal regions.'[6]

The next summer in Arctic Red River Henderson gave a demonstration of radio telephony, using his two receivers, speaking into the ear phones, and communicating between two adjacent buildings. Radio of any description was a novelty, and the transmission of voice was clearly a wonder of the modern age. Then, on their homeward journey up the MacKenzie River on the *Lady Mackworth* Henderson set up a receiver and delighted all on board by copying for them the news of the world as it came over in Morse code. Captain Gardiner proudly proclaimed that his ship was the 'first Canadian steamer equipped with wireless in the Arctic Circle!' At several of the stations the observatory party left a memento in the form of a sundial, designed by Henderson and mounted on the cement pier built for Swinburn's transit instrument; in each case it was the community's only indication of the time of day [Plate 29].

The success of these two seasons in the north established Henderson as an authority on the time signals that might be useful and also on the radio equipment best suited for field work. He built several long-wave receivers for use by observatory and by topographic survey personnel, as well as a miniaturized long-wave receiver which would fit into a vest pocket.[7]

In some of his receivers Henderson anticipated a future innovation in radio manufacturing: band switching. 'Some years ago we put out the first of a type of combination set for operation on two specified wave bands, a double-throw double-pole switch changing from one band to the other ... Most of these sets have been made to cover bands 200–750 metres, and 7,000–20,000 metres, the shorter waves being used for concerts and also ships and weather, and the longer for time signals and long wave press reports. Any two bands could have been selected and built into the set.'[8] It was suggested to Henderson that he apply for a patent, but he declined.

The time service at Ottawa had three objectives: the determination of fundamental time from star observations, the wireless reception of time signals in order to compare the results of timekeeping at different observatories, and the dissemination of time locally and nationally for the benefit of all sections of the community. Only a few modern observatories were equipped to maintain a time service, a growing number pressing on

6 Otto Klotz, *J. Roy. Astron. Soc. Can.*, 15 (1921) 307
7 Personal interview with J.P. Henderson.
8 J.P. Henderson, 'Wireless time signals'

with other avenues of astronomical research and relying for their time on radio time signals emanating from those few. Prior to the advent of radio the local observatory had had some degree of confidence that the results of its own transit observations gave a time that was entirely adequate and accurate. Radio dispelled that myth. Lack of a standard star catalogue, or of a uniform method of reducing to mean solar time, led to discrepancies of several hundredths of a second. It was to study these differences that in 1918 the Dominion Observatory commenced the daily reception of wireless time signals. At that time Annapolis or Arlington was received; in April 1921 the Bordeaux (France) signals of three hundred beats were added to the program of reception, in August 1924 the Nauen (Germany) signals, and in January 1928 the signals from Rugby (England).

Originally the comparison was made merely by observing the coincidence between the audible beats of the incoming signal and the local chronometer. Some degree of precision could be obtained, but there was liable to be a personal equation with different observers. After 1920 the method of wireless reception employed was 'coincidence of signals by extinction,' using an auxiliary clock. The auxiliary clock or chronometer was equipped with a contact which was open half a second and closed half a second, so that the output of the radio was cut off for the first half of each second. If the auxiliary clock had a gaining rate, its beats, which caused blanks, would gradually overtake the radio signal, nipping off a little more each second, and finally obliterating it, and the moment of coincidence was judged to be the clock beat that just completely blanked the radio signal. In the case of a vernier time signal which beat sixty-one times to the minute, a sidereal chronometer replaced the gaining chronometer. The radio time signal, which had a faster beat, came earlier and earlier and was gradually hidden in the 'open' caused by the chronometer. Then a few seconds later it reappeared at the leading edge of the break. The moment of coincidence in this case was just at the first reappearing pip of the radio time signal. The coincidence in each case was estimated to the nearest half second.

The favourite location for the chronometer, or for the relay operated by the chronometer, was at the ground side of the antenna circuit. It was reasonably effective in removing the incoming signal from the earphones during the half-second 'opens.' A new idea was conceived in 1925 when Henderson placed the chronometer between two matching step-down transformers back-to-back on the output of a receiver, in such a way as to provide makes and breaks which were electrically clean and noise free and were no strain on the delicate contact points in the chronometer. From then on, all radio sets both for field or for laboratory time signal duty were fitted with the back-to-back transformers for coincidence comparisons. The output of the receiver was not disrupted, and in fact was only partly used. In the event of good, clear reception it was then possible to tap in ahead of the transformers and record the signal on the drum chronograph as well as compare it by the coincidence technique.

The US naval system of signal transmission seems always to have been based on sixty beats to the mean time minute, except for omissions of the 29th and the 56th to 59th beats each minute. The last minute of the five-minute transmission omitted the last ten beats, 50th to 59th. For mariners, the ship's chronometer could be set to the nearest second by visual monitoring while listening to the radio time signals. A more exact comparison for scientific purposes required a gaining chronometer, which at first was set to gain about fifteen minutes a day, thus assuring a coincidence once in about ninety seconds or three during the five-minute interval. A difference of a second in the determination of the coincidence meant a difference of one hundredth of a second in the final clock correction. The drawback to this method was that one coincidence might occur during the five-second silent period, and a surge of static could make one or both of the others uncertain. An increase in the gaining rate to about thirty minutes a day resulted in a coincidence every fifty beats, or five during the five-minute transmission interval. By estimating the moment of coincidence to the half second there was but small drop in accuracy, compensated for by the increased number of estimations.

The European stations made extensive use of a transmission involving sixty-one beats to the mean time minute, in which the minute markers were lengthened for identification. A coincidence with a mean time chronometer was assured once each minute, by means of which the mariner could readily check his chronometer to within a few sixtieths of a second. Reception of this type of signal at Ottawa was readily accomplished using a sidereal chronometer, the coincidences occurring about seventy seconds apart for a total of four or five during the five-minute transmission period. In both cases the comparison chronometer was carefully compared with the primary clock before and after the signal reception so that the time of the signal was ultimately known in terms of the corrected astronomically determined time of the observatory.

Longitude would be improved by extending measurements over long arcs covered by the radio time signals. By 1919 there had been 174 longitude stations established in Canada, fourteen of them by wireless, and R.M. Stewart, at that time head of the Time Service, proceeded to adjust the net of Canadian longitudes. The solution showed a disagreement of 0.123 second between the difference of longitude of Montreal and Seattle depending on whether it was carried through the Canadian or the American net. Since the longitude between Seattle and Kamloops was of very small weight it was decided to use only the tie to Harvard College and to disregard the Seattle-Kamloops arc. The solution gave Ottawa and Vancouver the longitudes 5^h 02^m $51^s.940$ and 8^h 12^m $28^s.348$ respectively. The adopted longitude of Ottawa formerly had been 5^h 02^m $51^s.983$.

After much preliminary discussion, carried on over several years, it was decided at a meeting of the International Astronomical Union at Cambridge, Massachusetts, in July 1925 to carry out an extensive scheme of longitude determinations in October and November 1926. Almost a quarter of a century had elapsed since the transpacific longitudes had been measured. During this interval advances had been made in technology. Emission and reception of wireless time signals had replaced the telegraph wire and the submarine cable with its repeaters, and measurements were being made with increasing accuracy. Now it was appropriate to measure long arcs extending from continent to continent and encircling the globe. 'The test of the Wegener theory of continental drift was admittedly the most active stimulant to the desire for such measurements.'

'Forty-two different observatories and observing stations covering almost all of the inhabited parts of the globe took part in the measurements. Canada occupied two stations, one at the Dominion Observatory and one at Brockton Point, near Vancouver, this being the point from which the transpacific longitudes were started. As finally adopted, Algiers, Zikawei and San Diego were accepted as one main figure, and Greenwich, Tokyo, Vancouver and Ottawa as a second principal figure. The longitude of the other points were deduced differentially from these points.'

At Vancouver the observatory at Brockton Point was enlarged, and an insulated clock room was added to house the Howard sidereal clock and a half-second gravity pendulum. C.C. Smith and H.S. Swinburn operated a straight transit (Cooke) and a broken transit (Heyde) respectively, the Cooke being 12 feet 6 inches to the east of the Heyde. J.P. Henderson set up the aerials and receiving equipment and connected a double line to the home of J.H. Walsh, keeper of the Brockton Point time gun, to provide him with time signals, concert music, and an intercom with the observatory. The radio equipment included a double-wave receiver, a short-wave receiver, batteries, charger, and spare parts. A.H. Miller, who had gone down the Mackenzie River with Swinburn and Henderson in 1921 and 1922, took advantage of the timing facilities provided by the observatory at Brockton Point and installed his gravity pendulum. It was a precise half-second pendulum which, according to Henderson, served not only as a means of measuring gravity, but additionally as a check on the Howard clock.

A near tragedy almost overtook the Vancouver group. C.C. Smith, who had spent several years as a surveyor at the west coast, went on an inspection flight on 25 October 1926. The glassy calm of the water on the return of the flight proved deceptive, causing the plane to crash. The occupants were spilled into the water. Smith surfaced within the radius of the still-rotating propellor, but fortunately escaped with only a scalping. After two weeks of hospitalization he was back on the job, making daily visits to the doctor for dressing. Five or six years were to elapse before the head wound finally closed over, but it did not detract from Smith's ability and drive as chief of Canada's time service.

Computations conducted at Ottawa and reported to the American Astronomical Society at Amherst, Massachusetts, in Sept. 1928 gave the adjusted results for the primary figure which included Greenwich, Tokyo, and Vancouver. The longitude of Ottawa was $5^h 02^m 51^s.928$, 0.012 second smaller than the value reported above for the 1920 adjusted results. For Vancouver the value was $9^h 12^m 28^s.343$, 0.005 second smaller. The weight attached to these new results was too small for any changes to be made.

In 1933 the whole program was repeated, with improved instrumentation. Smith had converted his Cooke transit to a broken type, so that he and Swinburn each observed with broken type transits, Swinburn using a Heyde. At Ottawa, also, W.S. McClenahan used a Heyde, while R.J. McDiarmid used the meridian circle. All four observers were thus able to use zenith stars for time purposes, thus minimizing the error due to azimuth. In 1933 a group of stations located at Cape of Good Hope, Adelaide, Wellington, and Buenos Aires formed a southern figure. Then to connect all observing locations, both north and south of the equator, a group of equatorial stars was observed by everyone. The same time of year, October to November, was selected, with the suggestion that principal stations such as Ottawa and Vancouver might start 15 September and continue till 15 December, which was done. In the 1926 campaign, observations had been conducted in the first half of the night and a second set in the second half of the night in order to co-ordinate efforts with those observing to the east and those observing to the west. Also, an independent check was thereby made possible of the systematic difference in the results of A.M. versus P.M. observations which some people had reported. Some changes were made in the 1933 campaign. Transit observations for time were taken before midnight local time, and a common star catalogue (Eichelberger) was used. Radio time signals were restricted to those in the long-wave band, because short-wave transmissions had certain anomalies of radiation not well understood.

The schedule of time signals (in Eastern Standard Time) followed at Ottawa was NSS (Washington) noon, GBR (Rugby) 1 P.M., FYL (Bordeaux) 3 P.M., NSS 4 P.M., NSS 10 P.M., NSS 12 midnight, NSS 3 A.M., FYL 3 A.M., GBR 5 A.M., DFY (Nauen, Germany) 7 A.M. Washington signals terminated on the hour, while the European signals (rhythmic) commenced one minute after the hour, both having a duration of five minutes. They were observed by the method of coincidences by extinction, and when strong and free from static were also recorded. The Ottawa program of radio time signals was handled by me, while J.P. Henderson followed a similar program at Vancouver [Plate 47].

Every effort was made to measure and account for errors. A personal equation machine was employed by the observers at both Ottawa and Vancouver, and careful note was made of instrumental constants and relay lag. At the conclusion of the campaign a record of all the radio observations in terms of the local sidereal time was forwarded to the Bureau International de l'Heure in Paris. The number of permanent observatories or astronomical stations participating in the second campaign was about eighty-seven, almost double the number in the first campaign. The interval of seven years was actually too short to give any evidence for or against the Wegener theory of continental drift, mainly because the instruments available were not sufficiently precise. C.C. Smith, in reporting to the Royal Society of Canada in May 1935, stated that though the evidence was inconclusive, the exercise was an outstanding demonstration of international co-operation. A valuable by-product was the more accurate determination of the longitude of participating stations, a comparison of the right ascensions of the stars used, and some indication of improvements in radio instrumentation.

At the close of the longitude campaign in 1933, wireless reception of time signals was extended into the short-wave band, the program in the table being observed on a seven-day-a-week schedule at the Dominion Observatory. The comparisons were by coincidences, and when the signals were strong enough they were also recorded on the drum chronograph.

During 1933 and 1934 nearly fifteen hundred wireless comparisons

Program of time signal comparisons, 1934

Station	GMT		kHz	
	h	m		
DFW (Germany)	16	06	16.55	
NAA (Washington)	17	00	8410.	(Jan 1938 NAA 113 kHz)
GBR (England)	18	00	16.0	
FYL (France)	20	06	15.7	
LSD (Rio de Janeiro)	23	50	8830.	
PPE (Monte Grande)	00	00	8721.	
NAA (Washington)	03	00	9050.	(Jan 1938 NAA 113 kHz)

NOTE: LSD observations were started in April 1933, PPE in March, 1934, NAA short wave in July 1934, when the long wave was discontinued, and DFW on 14 March 1935, when DFY was discontinued.

were made by the two methods, coincidences and recording, yielding an average difference $(T_c - T_r)$ of +0.014 second, which was ascribed to a lag in the recording circuit. The signals were for the most part from the relatively close US Naval Observatory, the European signals being recorded only a few times.

Henderson's long-wave receivers continued in general use in the wireless room and by government field parties during the 1930s, the field sets coming back regularly for maintenance and repair. I assembled a tuned radio-frequency short-wave receiver, which was useful because of the increasing emphasis on short-wave time signals. Generally, though, the research in receiver design subsided with the availability of commercially designed receivers.

The retirement of C.C. Smith as chief of the Time Service at the end of 1937 left a vacancy which was not filled because of the depression. The outbreak of war in 1939 and the further reduction of staff due to war service made it necessary to curtail activities and to leave the time room unattended for longer intervals. Henderson, who had experimented with five-metre transmission and reception for several years, arranged to monitor two of the time service functions using a 1000-hertz and a 400-hertz tone modulating two five-metre transmitters at the observatory, and these tones were received at his home about two miles distant. One provided an alarm in the case of a failure of seconds pulses to the CHU transmitter site; the other monitored the circuit to the broadcast station CBO. Hollinsworth also provided himself with a five-metre receiver in order to collaborate in the monitoring program.

The observatory had become increasingly noisy electrically with the increase in clock circuits and the development of transmitters. The latter had commenced as five-watt units, except for one or two intervals each day when a more powerful tube was turned on to radiate on the 40.8 metre channel. By the outbreak of war the three frequencies were operating continuously with inputs ranging up to fifty watts. On 7 June 1944 a private telephone line was installed between Henderson's home and the observatory. The receivers were moved to the quiet location of his home. A stepping switch which could be controlled from the observatory selected in succession NSS (Naval Observatory, Washington) on 113 kHz, WWV (Bureau of Standards, Washington) on 5000 kHz, VAA (Department of Transport, Ottawa) on 8330 kHz, CBO (Ottawa) on 910 kHz, and CFH (Halifax) on 105 kHz.

The main source of electrical interference at the observatory ceased in 1947 when three RCA 300-watt transmitters were installed at the Department of Transport Short Wave Station on the Greenbank Road and CHU transmissions from the observatory ceased. Other noise-reducing steps had also been taken. Ordinary electric wall clocks, some operated by crystal-generated sixty-cycle, others from the ordinary wall plugs of the lighting circuit, replaced second and minute jumpers; mechanical relays were replaced by grid-operated tube relays, and pendulum clocks were finally replaced by crystal frequency standards. Relay clicks became a thing of the past.

By 1951 the attic room, as it was called, was free from all transmitting electronics. With a little rearrangement it became the home of a group of receivers, any one of which could be selected from the wireless room. The telephone line to Henderson's home was continued for another year or two for experimental purposes, then discontinued. GBR (England) had been restored to the list of sixty beats to the minute as well as the rhythmic signal of sixty-one beats to the minute. From a comparison of the two it was evident that they were derived from the same source, and monitoring of the rhythmic signal was discontinued.

Technology advanced rapidly in the post-war years. The method of reception using coincidence by extinction gave way to the method in which the radio signals were compared with microdial readings on a General Radio synchronometer. Several settings with the microdial were made to the point where the leading edge of the incoming signal was just obliterated by the microdial contact. Reading to a tenth of a division meant that the comparison was made to a nominal accuracy of a millisecond.

The oscilloscope made possible a visual display of the incoming signal, and the delay counter, which controlled the start of the sweep to the tenth of a millisecond, afforded a means of comparing the signal and the observatory reference clock with this precision. The method was particularly useful for monitoring the output of CHU, and frequently the more distant signals, especially WWV, were capable of reception with the same precision.

For a while the attempt was made to monitor the long-wave (very low frequency) seconds pulses of NBA, a US station at the Panama Canal. On the oscilloscope the slow build-up time which characterized these signals was clearly displayed. The static often made it difficult to determine where the signal started. By superimposing several seconds pulses from the oscilloscope onto a polaroid picture, it was possible to distinguish the actual signal shape because a repetition of at least part of the pulse each second made it the dominant feature, and its start could be determined. Subsequently low frequency signals proved to be more effective when using the carrier frequency instead of the seconds pulses as a vehicle for comparing time. Special tracking receivers, designed to accept both the incoming radio frequency and a frequency from the local standard, would register the difference, either in beats or in parts in 10^{10}, one part in 10^{10} corresponding to 8.64 microseconds in twenty-four hours.

Transmitters such as CHU were improved initially by good crystal oscillators. Then by 1960 a quartz crystal frequency standard was used to generate each of the three frequencies as well as the seconds pulses. A rubidium standard replaced the crystal, and a cesium standard replaced the rubidium during the 1960s. Time transmissions at other national laboratories followed a similar pattern of improvement.

In 1941, for the first time, a fundamental catalogue of stars, *FK3*, was accepted internationally. From this date forward transit observations for time were based on a unified system of star positions. Following the war, astronomical determinations for time were improved by an order of accuracy with the use of the photographic zenith tube, the Danjon astrolabe, and other advanced techniques. It became meaningful to compare clocks with earth rotation each clear night to a few milliseconds instead of a few hundredths of a second as formerly. Furthermore the introduction of small corrections to compensate for polar wander and annual fluctuation resulted in a time which had reasonably good short-term stability, though it suffered from small random fluctuations in earth rotation.

Because of improvements in the astronomical determination of time, and also in the broadcast and reception of time signals, the BIH was able to prepare a better international clock from all the current data. The contributions from the individual laboratories were then examined to see if there might by chance be a systematic difference. Any such difference was then interpreted as a small error in the adopted longitude. This was a global adjustment such as was hoped for in the world longitude campaigns of 1926 and 1933, but for which the technology of the day was not sufficiently advanced. Now, with much improved data, several small adjustments were suggested. The US Naval Observatory at Washington acknowledged a small adjustment at the beginning of 1961 and 1962. In a communication about this date from the chief of the BIH in Paris, the Dominion Observatory was informed that its adopted longitude appeared to require no change. R.M. Stewart, always a man of precision, would have been delighted to learn that the adjustment he made in 1920 had stood the test of time. More recently a 10 millisecond change has been made by the BIH to the Ottawa reference point, making it 5^h 02^m $51^s.950$.

By 1967, when the second was defined in terms of the cesium atom, commercially designed atomic clocks were available. One of them had been installed at the transmitter site to control the carrier frequencies and seconds pulses of CHU. The value of the transmissions was thus considerably enhanced, and the same was true of transmissions from other observatories.

For the intercomparison of the time maintained by the various laboratories around the world, the most important link has become the chain of Loran-C stations. Designed originally as an aid to marine navigation, the carefully coded emissions from each station are precisely controlled by cesium atomic clocks. The performance of each clock is checked by regular visits with a portable cesium standard. Special receivers designed to lock onto the code of the transmission provide a means of comparing both the time and frequency of the signal with the working standard of the laboratory. The time comparison remains relative until a visit to the laboratory by a portable cesium clock team establishes the travel time to the tenth of a microsecond from the Loran-C station.

One can readily recognize the change that has occurred with the introduction of the cesium standard. Originally the uniform marker against which time was checked was the mean solar day, and from it was derived the hour, the minute, and the second. Now the marker is the very short interval of time required for a transition to occur between two hyperfine levels in the ground state of the cesium atom. It corresponds to a frequency of 9, 192, 631, 770 cycles per second, and the main effort is to establish this frequency with increasing exactitude. By 1973 technology yielded a precision of a few parts in 10^{12}, which meant that if two standards could be kept in operation for one hundred years they would drift apart by only a few milliseconds. At the present rate of advance this precision should soon be improved by another few orders of accuracy, perhaps to a part in 10^{15}.

If the rise of radio time signals permitted the Dominion Observatory to attain unprecedented accuracy in timekeeping, the new technology also offered an opportunity to distribute time to Canada and the world much more efficiently than ever before. This side of the story began in 1923, when the first radio time signal was broadcast from the Dominion Observatory. It came about as a natural consequence of J.P. Henderson's enthusiasm as a radio experimenter, and must have had the full support of Klotz, the chief astronomer, and Stewart, who was then superintendent of the Time Service.

The short wave end of the radio spectrum below two hundred metres was essentially wide open to experimenters in the early 1920s. It was normal to take out a licence for each transmitter. Henderson acquired the licence 3AF in 1922 which had particular significance in that it was the sixth (F being the sixth letter in the alphabet) call sign issued for amateur experimental work in Ontario. In January 1923 he acquired 9CC for experimental work on 275 metres and 3VO for amateur communication. Later that year he and a few others of the Ottawa Amateur Radio association acquired the experimental broadcast license 10AP, which was operated from the wireless room of the Observatory. The ultimate objective of this station was to radiate time signals, but it was necessary first to have an acceptable program of music to which people would listen before introducing a short interval of seconds pulses.

Commander Edwards and Colonel Steele established a broadcast transmitter at the test room on Wellington Street with the call sign AP. Their chief interest was to test various wavelengths to discover which was the most effective in the spectrum below five hundred metres. For a while in 1923–4 there was a broadcast station in Ottawa operating with the call sign CHXC, and the test of a good receiver was its ability to tune out the local signals and pick up the more distant ones such as KDKA in Pittsburgh. Neither the receivers nor the transmitters were capable of sharp and accurate tuning. For this reason Henderson and his group suspended operation of 10PA in February 1924 when the CNR station CKCH was inaugurated in Ottawa; within a few weeks CKCH became CNRO, then CRCO, and finally CBO. A loop was extended from the CNR telegraph office to the studio of CNRO in order that every evening at 9 P.M. the Dominion Observatory time signal could be broadcast. Through the years the time signal broadcast was changed to 3 P.M. and then to its present location of 1 P.M. At first the pulses were formed by a buzzer. Within a matter of weeks an audio oscillator, tuned to 512 hertz, was installed in the studio. It proved to be so acceptable that a second oscillator was built for the broadcast of time signals from CNRA in Moncton NB, whose signals originated in the Saint John weather office.

During the summer of 1923 a seismological field station was set up at Shirleys Bay, about ten miles west of the Observatory. Transmitting and receiving facilities were established at the site so that the seismologist would be able to receive the time pulses from the observatory clock and also communicate directly with Henderson in the wireless room, the wavelengths used being mainly 225 and 275 metres. During the summer's operation both mean time and sidereal seconds pulses were transmitted to Shirleys Bay using the 9CC transmitter. Communication was maintained with the call signs 3AF and 3VO, which were regular amateur calls. 9OB was acquired sometime later for use in broadcasting time signals. The prefix VE, which today characterizes all Canadian amateur licences, did not become mandatory till 1928, when the large number of operators made it desirable to have a distinctly Canadian identification.

Many hours were spent by Henderson investigating the properties of short waves down to the five-metre range. His results formed the basis of some very popular demonstration lectures to the Ottawa Radio Association during the 1920s.

Time signals were emitted frequently, but on no fixed schedule. An attempt was made to determine how well they were received in different parts of the country; for this purpose Henderson maintained active contact with amateur operators, particularly in the Toronto-Hamilton area. Portable equipment was assembled which he carried in his car and canoe in order to test the effectiveness of the time signals and the field pattern of various transmissions in the Ottawa area.

In March 1923, WWV of the National Bureau of Standards in Washington DC (later Fort Collins, Colorado) commenced the valuable service of transmitting on various wavelengths and announcing the value of each to the exact metre. The first of these transmissions to be monitored by Henderson at the wireless room of the observatory, occurred on 29 May 1923. The station was monitored as the wavelength was shifted in ten-metre steps from 210 metres down to 150 metres, and the setting for each was carefully checked on the General Radio wavemeter. The voice announcements were not distinguishable, but they were not really required by the radio operator of that day.

On 1 March 1928 a regular daily transmission of radio time signals from the observatory was initiated. 9CC (now designated VE9CC) on 52.5 metres was on the air for five minutes from 2:55 to 3:00 P.M. The tube was a De Forest, operating with a plate input of about 1/4 kilowatt (150 milliamps and 1500 volts). Occasional transmissions were also sent after 10 P.M., and seconds beats from the sidereal clock as well as the mean time clock were placed on the air.

By now most of the survey parties were equipped with receivers to pick up the long wave time transmissions from Washington and elsewhere. There were also some field men whose operations were adjacent to the telegraph line, and for whom the old familiar method of exchanging time, using the wire connection back to the Dominion Observatory, was less complicated. But it was much more expensive. In J.P. Henderson's diary one finds the following notations: 'CPR telegraph to Saskatchewan border $20 per exchange – tonight 2 exchanges $40' (16 July 1925); 'Mr. McDiarmid was quoted as saying telegraph last year cost $1000 for 30 nights – $30 per night' (5 July 1928). Doubtless this justified the cost of developing radio equipment which would ultimately make time available everywhere in Canada.

The observatory short-wave transmitter continued to send time from 2:55 to 3:00 P.M. on a wavelength of 52.5 metres (later 52.6 metres) till 15 January 1929, when the wavelength was changed to 40.8 metres and the call letters to VE9OB. To make the pulses a simple Hartley oscillator circuit was used [Plate 48]. The key closed an absorption loop which was placed close to the tank coil, effectively changing the tuning, so that the signal had a 'front' and a 'back' wave. Correct tuning was indicated by a simple wavemeter consisting of a coil and condenser and a small series lamp that would glow when the wavemeter was in resonance with the transmitter. The transmitting antenna was strung between masts eighty feet or more high and about 350 feet apart. It consisted of two wires spaced about eight feet apart by spreaders at each end of the span. It was gratifying later that year to learn that scientists with the Geodetic Survey had been able to receive VE9OB while working north of Churchill on the 60th parallel.

During 1929, in addition to 40.8 metres, wavelengths of 80 and 102 metres were used experimentally. By the end of the year a five-watt transmitter on 90 metres, just outside the amateur band, was being operated continuously during the daytime. The supply of about ninety volts consisted of an array of Edison cells made up in test tubes held in a parafined lattice wooden box. A drop of oil in the top of each test tube prevented evaporation. The input power of five watts to the transmitter meant that the battery required recharging every two or three days.

With the acquisition of new types of equipment it became possible by 1933 to eliminate the batteries and maintain low-power transmitters on a 24-hour basis, the wavelengths now being 20.4, 40.8, and 90 metres. With condenser tuning, the stability was not good. Fortunately Ed Davey of the Radio Branch of the Department of Marine and Fisheries was now operating the newly established Monitoring Station from quarters on the Experimental Farm, not far from the observatory. His careful attention prevented the three VE9OB transmitters from deviating noticeably from their assigned wavelengths.

Quartz crystal oscillators for 40.8 and 90 metres were acquired in 1933. Crystals for 20 metres were available, but they were rather fragile, and it was general practice to double the frequency from a 40-metre crystal. A quartz crystal permitted the oscillator to be maintained within a very narrow tolerance of a fixed frequency. In addition, the output of a crystal-controlled transmitter had the desirable clear tone that closely resembled the pure 'dc' tone of a battery-operated tube oscillator. It was usual for a tube oscillator to include a portion of the ac ripple in its output when the high voltage was derived from a generator or a rectifier instead of a battery. If the crystal happened to be dirty or marred in some way, it could jump to a nearby frequency. In this event the first step was to clean it carefully in carbon tetrachloride to remove any dirt or grease. A chipped or cracked crystal, of course, would not be improved.

In 1934 the Department of Marine notified the observatory that the that the time transmission channels, as indicated in the Berne list, were 3332, 7350, and 14,700 kiloherts, corresponding to 90.00, 40.82, and 20.41 metres. This marked an official change in nomenclature from wavelength to frequency and an increasing emphasis on closer tolerances within the assigned frequencies. Since 1933 these tolerances had been 0.03 per cent for frequencies between 1500 and 6000 kHz, and 0.02 per cent for frequencies between 6000 and 30,000 kHz. Hence the assigned frequency of 7350 kHz meant 7350 ± 1.5 kHz.

The next change occurred in 1938 when CHU became the official designation of the Dominion Observatory time transmissions. In January

1937 the frequency assignments were modified to 3330, 7335, and 14,700 kHz and by the end of 1938 the high frequency channel was modified to 14,670 kHz, the exact harmonic of 7335 kHz.

Following the example of wwv, a tone of 440 cycles from a tuning fork was imposed onto the seconds pulses. This was later changed to 1000 cycles derived from the quartz crystal frequency standard. There was no phase relation between the pulses and the superimposed tone, because they were derived from two independent sources. The pulses were controlled initially by a pendulum clock, and after 1938 were produced by the time machines built by R.M. Stewart. The ability to count down to one per second and generate time pulses directly from a standard frequency was a post-war development and was not applied to CHU till the 1950s. The superimposed 1000-cycle tone had two purposes; it was of assistance in finding the signal, and it was accurate to a part in 10^7 for anyone who could use it as a calibration frequency.

The little transmitters were built and set in operation in the wireless room, and each fed by a single wire to one of the three horizontal antennas located on the roof. By the end of the 1930s each frequency fed into a tuned half-wave dipole with a twisted pair of ordinary lamp cord which terminated at the antenna with a V symmetrically flared about the centre of the antenna and with a pick-up loop at the output tank coil of the transmitter. During 1941 the three transmitters were removed from the wireless room to the attic or midway room of the observatory.

In 1934 the Dominion Observatory adopted the method employed by the US Naval Observatory for identifying continuous signals. At first a tape was used, but after 1938 the time signal machine provided the omissions by which identification was accomplished. The series repeated itself each five minutes as follows: 1st minute – 29, 51, 56-9; 2nd minute – 29, 52, 56-9; 3rd minute – 29, 53, 56-9; 4th minute – 29, 54, 56-9; 5th minute – 29, 51-9. Hence, the omission of the fifty-first second and then four more beats indicated four more minutes to a five-minute interval; at the end of the second minute fifty-two was omitted and three more beats sent, indicating three more minutes to the five-minute interval; and so on. The end of the fifth minute had the long gap from the fifty-first to the fifty-ninth beats. During the first minute of each hour, the call sign CHU was sent slowly in Morse code twice. The second and subsequent minutes of the hour corresponded to the second and subsequent minutes of a five-minute group.[9]

9 M.M. Thomson, 'Canada's time service'

The report by surveyors that they had heard the observatory time signal while on field duty north of Churchill during the summer of 1929 encouraged the optimistic hope that sometime in the future Canadian surveyors, wherever they might be in Canada, would be able to rely on time from the Dominion Observatory. The 10 P.M. signal that had been heard had had a power input of about 250 watts derived from a motor generator. About 1931 this transmission was discontinued. Thereafter it was necessary to rely on the gradual improvement in the power and efficiency of the three low-power CHU transmitters. During the depression years it was a slow process, and in fact might not have advanced at all had it not been for Henderson's ability to pick up bits and pieces of equipment at bargain prices and generously make them available for general use. By 1938 it was noted that CHU had an output of about ten watts, scarcely enough for Canadian coverage. During the next few years this was improved to about fifty watts. Meanwhile a direct line was established between the Dominion Observatory and the Department of Transport transmitting site on the Greenbank Road which permitted time signals to be sent over VAA, on 11990 kHz, from 10:55 to 11:00 A.M. each day. The lag was measured as 0.010 second. With a power of two kilowatts much better coverage of the north was achieved, particularly in the direction of Churchill, towards which the antenna was oriented.

There were times when the survival of CHU hung on a slender balance. Critics suggested that the daily noon signal radiated by the CBC perhaps obviated the need for CHU. In October 1941, R.M. Stewart, by then Dominion astronomer, acknowledged Henderson's pioneering effort: 'Imitation is the sincerest flattery. We have just recently learned that the US Bureau of Standards has, within the last few months, begun a continuous broadcast of time signals such as we have been sending for a number of years ... the inception of our continuous broadcast, which was the first of its kind in existence, was largely due to the enthusiasm and insistence of Mr. J.P. Henderson of our staff. Without very much encouragement he built up our first transmitting equipment, partly from apparatus of his own, and put it into experimental operation. Since that time it has, of course, been improved very considerably, though the power is still quite low.'[10] The fact that shortly before, on 28 August, the time determined by the Dominion Observatory had been designated by order in council as Canada's official time greatly enhanced the importance of the observatory time service. And it provided the strongest argument for the ac-

10 RG 48, Dominion Observatories, Correspondence file, Public Archives of Canada

quisition of improved instrumentation for both the determination and the dissemination of correct time.

In 1942, VAA, the Department of Transport transmitter, commenced all-night transmission of the observatory time signals for the benefit of scientific field parties during the summer months. For this service the frequency of 5405 kHz was used in addition to the transmission each morning on 11,990 kHz. On 24 September 1943 the naval station CFH at Halifax commenced broadcasting observatory time pulses for five minutes at 10 A.M. and 10 P.M. Eastern Standard Time for the benefit of navigation on the North Atlantic. The pulses were sent by the radio teletype circuit from the Department of National Defence communications centre in Ottawa to Halifax and from there by cable to the CFH control centre at Albro Lake, just outside Dartmouth.

Since the inauguration of the observatory in 1905 there had been a wire connection in Ottawa to the GNW (later CNR) telegraph office over which clock beats were sent. A loop from this line fed the signals to the Canadian Radio Broadcasting Commission (later CBC) studio, the Bank of Canada, Ottawa Electric, each of whose streetcars was equipped with a clock, and Birk's Jewellers. Officially, the two telegraph companies distributed over their respective networks the time which originated from McGill. But with the removal of the transit hut from the campus in 1926 McGill had ceased to provide a fundamental time service, depending instead on the radio reception of time signals from Washington or Ottawa to maintain the local clock from which the signal emanated. Upon occasion the signal fed from the Dominion Observatory to the Ottawa telegraph office was relayed to Montreal for direct transmission to the Canadian network.

The time signal machine provided a means of supplying a program on a regular daily basis to each of the telegraph companies, using a separate coded signal for each. The Railway Association of Canada agreed to the proposal, presented by R.M. Stewart, that clock beats from the observatory be used, but stipulated that the signals should be relayed through the Montreal office. Accordingly, early in 1945 the service was established. This in turn made possible a daily time transmission of the CN signal from radio station VAP, at Churchill, Manitoba, on a frequency of 500 kHz. The lag of this transmission as monitored at Ottawa was 0.274 second.

In July 1946 the services originating in the observatory's time room were summarized by J.P. Henderson:

– CBC, networks broadcast for 40^s until 1 P.M. Eastern Time, standard or daylight saving, whichever prevails at Ottawa (about 50 stations).

- CPR telegraphs 11:54–11:56 EST
- CNR telegraphs 10:58–11:00 EST
- VAP, Churchill, Man., 10:58–11:00 EST broadcast on 500 kHz
- CFH, Halifax, 1500 & 0300 GMT broadcast on 105 kHz, 9040 kHz, and 5502.5 kHz
- VAA, Ottawa, 10:55 to 11:00 EST broadcast on 11990 kHz
- CHU, Ottawa, continuous through 24 hours a day, broadcast on 3300, 7335, and 14,670 kHz
- VAA, Ottawa, dusk to dawn during summer months, broadcast on 5405 kHz
- Telegraphs through CBO to carrier circuits, Eastern repeater network 7:45–7:50 EWT (Eastern War Time)
- Synchronization of Government clock systems, Bank of Canada, Ottawa Electric, and Birk's Jewellers
- National Research, direct line carrying continuous time signals
- Frequency Monitoring, Dept of Transport, direct line carrying continuous time signals

By this time CHU was operating with a nominal power of fifty watts on each of the three frequencies. It was still quite inadequate for all parts of Canada, but the local radiation was sufficient to cause some interference on broadcast receivers near the observatory. An increase in power would therefore have to be accompanied by a relocation of the transmitters. Space was generously made available at the Department of Transport transmitter site on the Greenbank Road because W.B. Smith of the Radio Division of DOT was a strong supporter of every step that would improve the reception of CHU by frequency monitoring stations across Canada.

Application was made in November 1946 for the transfer of three RCA transmitters, known to be war surplus, from the RCAF to the observatory. Support to the application was given by the Canadian Radio Wave Propagation Committee, for whom a strong continuous signal would be useful for ionosphere, auroral zone, and other research. As a consequence the application was granted. The three transmitters were installed and went into operation on 21 June 1947. Each had a rating of 300 watts, a sixfold increase in power for CHU. Simple dipole antennas were mounted between existing masts which had originally supported rhombic antennas [Plate 59]. Hence there was no provision for selecting the optimum direction for the radiation pattern. Keying was done from the time room of the observatory and consisted, as before, of pulses of 1000 cycles of approximately one-third of a second duration. At first, maintenance of the transmitters was performed on a 24-hour basis by Department of

Transport technicians living at the site. Subsequently the station became unmanned, and an alarm was installed at the air radio console of Uplands Airport where it would have around-the-clock attention. The transmitters, together with all the spare parts required for their maintenance, remained the responsibility of the observatory.

Soon after C.S. Beals assumed the duties of Dominion astronomer in 1946, he suggested to McClenahan that one large transmitter be acquired and placed on the frequency that would offer the best Canadian coverage, namely 7335 kHz. In 1951 Treasury Board granted $20,000 for a Collins transmitter designed to give a continuous output of three kilowatts. The need to enlarge the building led Beals to assign $10,000 of observatory funds to the cause. Finally, in January 1953, the new transmitter was placed in service, broadcasting on 7335 kHz, the older one being retained as standby. At first the new transmitter was maintained at full load only during the day time, being reduced to half load at night. In spring 1954 the radio teletype time signal from Ottawa to Albro Lake, the control centre for CFH at Dartmouth, was discontinued; the following August an attempt was made to key the transmitter directly from the CHU radio pulses, using the new 7335 kHz signal, but vagaries of reception, static, and other overriding transmissions made the experiment short-lived.

In these years there were also many changes in personnel at the Time Service. The radio pioneer, J.P. Henderson, retired in March 1956, after nearly thirty-seven years. He had kept the observatory abreast of the latest developments in electronics, and was deeply appreciated for his skill and patience as an experimenter and instructor. The next year, in July 1957, Bill McClenahan retired after forty-three years, nearly fourteen of them as head of the Positional Astronomy division. He had been responsible for publishing all the early observational records of the meridian circle transit telescope, had piloted the PZT through its early years, and had seen the time service finally abandon the pendulum in favour of the quartz crystal oscillator. He was a capable man and well liked by his staff. When he retired, I succeeded him as head of the Time Service and saw the introduction of the atomic clock into Canadian timekeeping and the adoption of the atomic second. When Canadian government astronomy was consolidated under the jurisdiction of the National Research Council, I saw the formation of the Time and Frequency section in the Physics Division of NRC, and the time laboratory became the nerve centre of Canada's time service [Plate 57]. In March 1972 I relinquished my administrative duties to C.C. Costain, and retired from the government service at the end of the year.

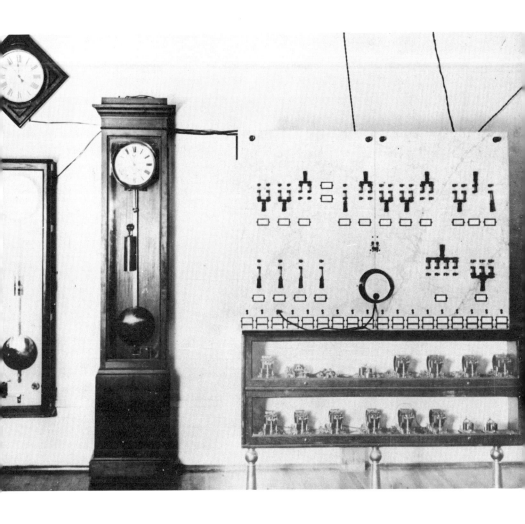

43 / The time room of the Dominion Observatory soon after it was established in 1905. Each secondary pendulum, sidereal on the left, mean time in the centre, has an electromagnet in the case and a soft iron bar at the bottom of the pendulum to keep it in synchronism with its primary pendulum in the clock room. The marble control panel is on the right, and below it are the relays which distributed the time pulses.

44 / LEFT A secondary clock showing the synchronizing system introduced by R.M. Stewart in 1913. At the upper end of the pendulum is a tray. Two small weights on light chains may be lowered onto the tray or held suspended to make the clock gain or lose. When power failed, one weight would ride and the other would be held up, and in this condition the pendulum was adjusted to maintain the correct rate.

45 / RIGHT Synchronome master pendulum s29, installed in the new clock vault in 1930, was the primary time standard at the Dominion Observatory for two decades.

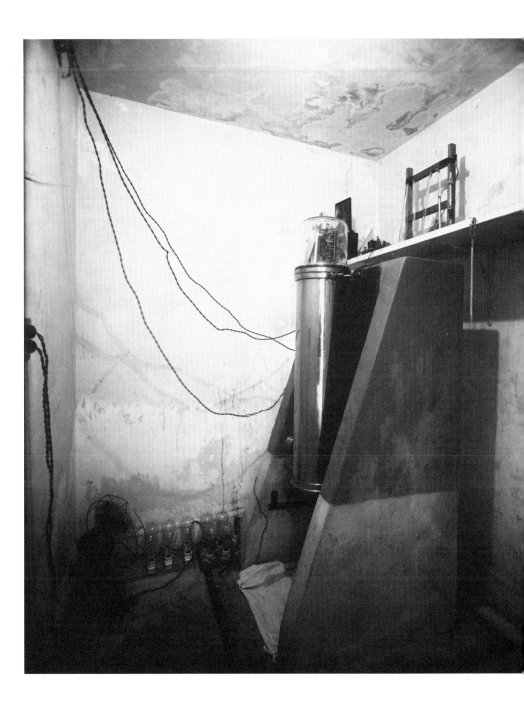

46 / J. Hector, watchmaker at the Dominion Observatory from the mid-1930s till 1948. The most modern facilities, together with sidereal and mean time dials, were used to check and regulate the field chronometers. Organized in the early days of the observatory, the watch repair shop was transferred to the Geodetic Survey Branch in the 1960s.

47 / The wireless receiving equipment at Brockton Point observing site,
Vancouver, 1933. Several chronometers, receivers, and a drum chronograph may
be recognized.

VE 90B. 40.8 m.

made up several coils of (5/16"?) Cu tubing to fit large REL V.C. make good tank circuits of low Resistance.

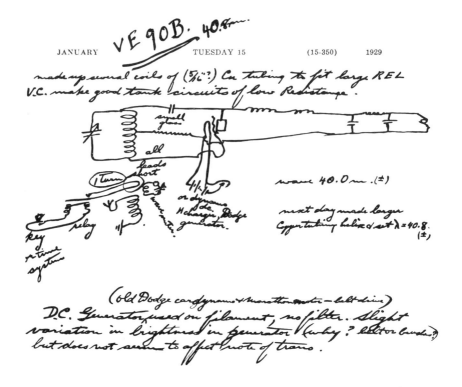

wave 40.0 m. (±)

next day made larger Copper tubing helix & set λ = 40.8. (±)

(old Dodge car dynamo + mon. ... motor — belt drive)

D.C. Generator used on filament, no filter. Slight variation in brightness in generator (ruby? ... ?) but does not seem to affect note of trans.

48 / Page from the diary of J.P. Henderson showing a diagram of a Hartley oscillator used as a transmitter

49 / J.P. Henderson in the wireless room, Dominion Observatory, 1949
Photo by Malak

50 / TOP LEFT J.P. Henderson adjusting the phase of one of the two Muirhead timing devices installed in 1952

51 / LEFT The author at one of the Western Electric primary frequency standards in the clock vault in 1958 where formerly an astronomical pendulum had been mounted. The vault provided both thermal control and freedom from vibration due to street traffic.

52 / BELOW A.G. Mungall, H. Daams, and R. Bailey observe the point of maximum resonance of the cesium beam as the input frequency is varied.

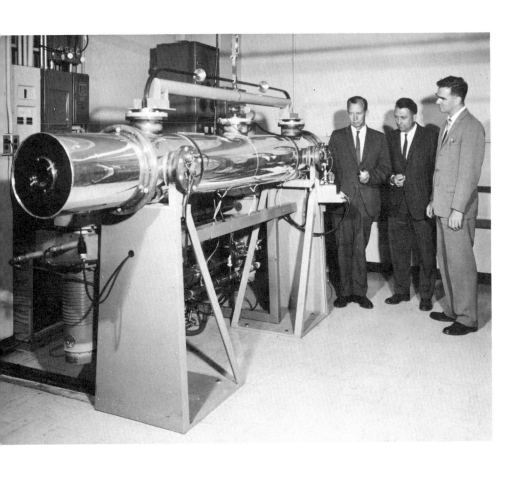

53 / Cesium III, the NRC primary frequency standard since 1965. R. Bailey, H. Daams, and A.G. Mungall were responsible for the design and construction.

54 / Cesium V, similar in design to Cesium III, approaching completion. J.Y.J.
Lauriault, H. Lam, A.G. Mungall, and H. Daams participated.

55 / Swiss atomic standard, en route to Expo 67, was brought to the time laboratory of the Dominion Observatory. Left to right: M.M. Thomson, Helmut Brandenberg, Peter Kartaschoff, V.E. Hollinsworth, and L.G. Miller

56 / A.G. Mungall and D. Morris examine the glass bulb of a hydrogen maser at NRC.

57 / The time laboratory of NRC is the nerve centre of Canada's time service.
Official atomic time is determined and distributed, and is compared by radio and
TV methods with the atomic time of other laboratories.

58 / V.E. Hollinsworth adjusting the talking clock built by Ateliers Brillié Frères of Paris. Photoelectric readers scanned the voice strips which were wrapped around the drum in grooves, one for each of the twenty-four hours and sixty minutes, and two for the preamble.

59 / Aerial view of CHU transmitter site in 1960, showing the array of rhombic antennas used by the Department of Transport. A few poles near the building supported the CHU horizontal dipoles. At lower left is the sterba curtain with a long feeder line discontinued in 1968.

Another advance in time distribution was the so-called speaking clock. When the American code of time signal identification had been adopted for the Canadian time transmission, it was assumed that a properly equipped surveyor would in all probability have a timepiece which would indicate the correct time within a minute or so. The simple code would then indicate to him which minute of a five-minute interval he was listening to. It turned out that code, no matter how simple, was unacceptable when an easier method became available: voice announcement. Voice had largely replaced code in wireless communication by 1952. Henderson had recommended that a rental contract be negotiated with a manufacturer, Audichron of Atlanta, Georgia. Departmental policy at the time, however, was opposed to the binding effects of long-term rental agreements, preferring instead to make an outright purchase if possible.

An examination was made of several voice announcement systems, both in North America and Europe. There were also discussions with people from NRC and from Crawley Films, Ottawa, about the possibility of a locally designed system. The outcome of it all was that Ateliers Brillié Frères of Paris were awarded a contract to supply two of their standard units, which appeared to combine ruggedness and fidelity at a price that was within reach [Plate 58]. It was stipulated in the agreement that the announcements be made by a male voice with a Canadian accent. Frederick Martyn Meach of the Canadian Embassy in Paris was seconded by the company to make the recording, and for a few years thereafter his voice was on the Canadian airwaves more than that of any other Canadian. The speaking clocks were housed in the basement clock room from which the piers had been removed. The voice announcement each minute of the twenty-four hours commenced in December 1954. It took the form 'Dominion Observatory, Canada, Eastern Standard Time, fourteen hours, forty-five minutes' and was in English only. On the hour the announcement took the form 'fifteen hours exactly.'

The recording for the Ateliers Brillié Frères speaking clock was made as a sound track on photographic film. The individual strips, one for each of the sixty minutes and one for each of the twenty-four hours, were wrapped around two drums in grooves. Light from the exciter lamp penetrated the film and was reflected back through the film to the photo cell by the reflective coating of the drum. Photo cell and exciter lamp were mounted on a unit called a reader which was moved mechanically along a track in discrete steps so that it faced successive grooves. There were three readers, one for the preamble, another for the hours, and a third for the minutes. The very short life of four to six weeks for the exciter lamps was of concern until the photo cells were replaced by lead sulphide cells, which were sensitive to red light. Then, glowing at only half illumination, the exciter lamps had an indefinite life.

The very nature of double passage of the light beam through the sound track film seemed to cause a reduction in fidelity. The announcements were sufficiently intelligible when monitored right at the speaking clock. By the time they had traversed the various circuits to the transmitter and returned as a radio signal there was a further drop in fidelity which occasionally involved monitoring two or three successive announcements to be sure of the correct time. The problem was aggravated, of course, as reception conditions deteriorated at greater distances.

When the change in speaking clock was made some six years later, several changes were incorporated. Perhaps the most important was the change from purchase to rental of equipment. Audichron was invited to draw up a contract for the rental of one of their speaking clocks; they would service it regularly and in addition respond immediately to an emergency call. With the expectation of such good service it did not seem necessary to rent a duplicate clock as standby. The high quality of the Audichron clock and the effectiveness of the regular service calls in fact justified this decision. The recording was done by Harry Mannis of the CBC and was transferred from magnetic tape to magnetic drum, preserving a high order of fidelity. The long line of communication from the observatory to the transmitter site was made obsolete by installing the speaking clock in a room adjacent to the transmitter. Commencing May 1960, therefore, the voice of Harry Mannis announced the time via CHU. The older equipment was retained intact in the event of a breakdown, which fortunately never occurred.

The old speaking clocks in the basement clock room did not become idle with this change. One of them had been adapted to an automatic time announcement over the Ottawa telephone system. Preliminary discussions of a time service of this nature had commenced in 1955; in December 1957 a formal request for an unlisted telephone line was made, supported by a count V.E. Hollinsworth had made showing as many as 444 calls for a time in one week, eighty-nine in one day, with a peak of twenty-two in thirty minutes, involving much of the attention of one person. Treasury Board at first turned down the request; Dr Beals stated strongly that the distribution of time by the observatory was an obligation to the Canadian public. Treasury Board, reconsidering, granted the installation of an unlisted line to the time laboratory by Bell Telephone. In January 1959, the automatic answering service was installed. It provided a

ten-second listening period to the talking clock, during which two complete announcements of the form 'five hours, five minutes' would be
heard. Entry to the talking clock was random, so that any part of the
announcement might be heard first. Twelve-hour notation instead of
twenty-four was used, and the time was advanced one hour during the
summer daylight saving period. The telephone number soon became
generally known, and the daily count of calls rose steadily. Interestingly
enough the number of calls during the night seemed to equal those
during the day. A private line of this type was considered by Bell Telephone to be fully loaded when handling about eight hundred calls in a
24-hour period; but by 1967 there were occasions when the daily count on
the automatic answering line exceeded three thousand. Bell Telephone
was unhappy with the situation, and there was a growing volume of
complaints from the public. One solution, requiring additional lines and a
French as well as an English announcement, involved a tariff scale by Bell
as well as technical arrangements that were quite unacceptable to the
observatory. The other solution was to discontinue the service, which was
done in March 1968. There were complaints. Yet public feeling in Ottawa
might have been more vocal had there not been alternatives. There was,
for instance, the 1 P.M. signal over the CBC network, not to mention the
increasing use of inexpensive multiband transistorized receivers that
could readily be tuned to one or other of the CHU frequencies.

The demand for easy access to a time signal was resulting in companies
producing radio receivers with pre-tuned components, so that a turn of a
switch would select a WWV or a CHU time signal. For more local use, a few
fixed-frequency receivers were assembled by V.E. Hollinsworth and S.H.
Sheard at the observatory. One of these was made available to the gunner
responsible for firing the noonday gun from Majors Hill Park, and
another to a government office responsible for the calling of tenders on
public projects.

The new speaking clock, whose announcements showed the advantages of improved technology, attracted the attention of amateurs and
short wave listeners, and brought requests for QSL (acknowledgment)
cards from many parts of the world. One Canadian listener, A.G. Williams of Ville La Salle, Quebec, wrote proposing that the official time
announcement of the Canadian government would be considerably enhanced if it were in both French and English. There were no funds
available in the observatory vote to make such a change, but this reply only
spurred Williams on to more action. In the fall of 1962 he addressed a
letter to the Hon. Paul Comtois, the minister of mines and technical

surveys, of which the Dominion observatory was then a Branch, and the following year he sent a further memorandum to the new minister, the Hon. William Benidickson. The reply this time was that the matter was under active review. Before long, with departmental approval, a new contract was entered into with Audichron. Harry Mannis again provided the English recording and the late Miville Couture of CBC Montreal, the French, giving an announcement of the form 'CHU Canada, Eastern Standard Time, fourteen hours, forty-four minutes, quatorze heures, quarante-quatre minutes,' followed the next minute by 'CHU Canada, heure normale de l'est, quatorze heures, quarante-cinq minutes, fourteen hours, forty-five minutes.' The 24-hour notation was used in order to avoid any ambiguity with respect to A.M. and P.M. or noon and midnight. The bilingual announcement commenced on 1 April 1964.

A new need arose for a local signal involving more frequent voice announcements and fewer pulses. It gave rise to a format involving six announcements per minute, followed by a pulse every tenth second. An additional lease agreement with Audichron in 1969 involved two complete sets of announcements, one in English and one in French, covering every tenth second in the twenty-four hours. Three possibilities are available on the output of the new speaking clock. The announcements may be alternately English and French, all English, or all French, with simultaneous output of all three options. Harry Mannis and Claude Miguet, both of CBC Toronto, provided the announcements, which take the form 'NRC, Eastern Standard Time, fourteen hours forty-four minutes, twenty seconds, BEEP.' The BEEP is a short burst of about 200 cycles of a 1000-cycle tone, and is accurate in time to about a millisecond. On 1 January 1977, a new telephone answering service was established in Ottawa using the new ten-second announcement and the BEEP. Two separate lines were used, one for the French, the other for the English announcement, with multiple access on each line.

As soon as CHU advanced from a purely local phenomenon to a service with increased output that could be extended to the wider Canadian community, two objectives were proposed: that the frequencies of the three channels be synthesized from a frequency standard and that the signal pulses be made uniform by the same source. When the keying had been controlled by the Stewart time signal machine the long-term value of the time was fairly uniform, but there were short-period variations. The correction imposed upon the small control pendulum by the primary clock, S29, had caused the length of the seconds to be shorter or longer than normal by a few ten-thousandths of a second during the brief interval of advance or retard. The total correction then could change the time by a hundredth of a second or more. It was possible for variations of this sort to occur every six minutes, which is the coincidence interval between mean time seconds and sidereal time seconds. The 1000 cycle tone superimposed on the seconds pulses was derived from a separate source, tuning fork or frequency standard, and hence bore no phase relationship to the pulses or to the carrier frequency.

When the Muirhead time machines replaced the pendulum controlled time machines in 1951 it became possible to adjust the phase of the pulses so that they commenced with a 1000-cycle wave peak, since both were controlled from the same 1000 cycle source. The transmitters were still dependent upon commercially designed crystals for frequency stability, so that there was no phase relationship between carrier frequency and signal pulse. If they should be in phase, then, in the case of the 7335 transmission, for instance, successive pulses would commence when the carrier had completed exactly 7335 thousand cycles, and each wave of the 1000 cycle tone would be spaced exactly 7335 cycles. So the leading edge of each second would coincide with the leading edge of a 1000 cycle wave as well as with the leading edge of a carrier frequency wave, and the result would be a series of seconds pulses that were both positive and extremely accurate.

Several groups were interested in this improvement, including the Department of Transport, which was responsible for issuing radio licences and for monitoring the frequencies of transmitters, and NRC, which was responsible for research into primary frequency standards and time, and relied upon a source of accurate time for much of the research. W.B. Smith, who had in post-war years become engineer in charge of the Department of Transport radio monitoring facilities, supported the move to increase the range of CHU time transmissions, and had advocated as early as 1948 that the frequencies be derived from a quartz crystal frequency standard. The Ottawa monitoring station as well as other monitoring stations within the range of CHU could then base all frequency measurements on a Canadian standard derived from Canadian observations of earth rotation. Smith collaborated in every way possible by providing space and technician services at the Greenbank Road transmitter site.

Specifications for a device that would bring into phase the frequencies and pulses of CHU were submitted to several companies, and three replies were received, the lowest tender being nearly three times the amount DOT was prepared to pay. The observatory and NRC were prepared to provide additional sums, but not to that extent. On 3 July 1958, a committee of representatives from DOT, NRC, and the observatory time service decided to continue with the project study and attempt to reduce the cost without sacrificing any features of the specifications. The result turned out to be far better than anticipated. J.C. Swail, a blind electronic engineer who worked in the radio and electrical division of NRC, proposed that they do the job themselves; in other words that they would engineer a system in which the carrier frequencies of CHU would be maintained at an accuracy and precision consistent with standard frequency transmissions, and the actual seconds pulses would bear a known and constant phase relationship to the carrier frequencies. So it was that using his own circuitry Swail produced the three CHU carrier frequencies from a 100 kHz Western Electric primary frequency standard, while from one of the Muirhead time machines the sequence of seconds pulses was arranged. Swail had the assistance of a competent technician, but his own genius extended to the point that he not only worked out the practical details from the theory but was also able to do some of his own soldering. The completed device was assembled and installed in 1959 at the transmitter site; making it a self-contained time source. Lines of communication from the transmitter site to the observatory permitted monitoring of both the control and the standby signal generators. A continuous strip chart on the output of three receivers provided a good record of the performance of each of the three transmitters. Early in 1961 a James Knights transistorized frequency standard replaced the older Western Electric unit. In 1964 Swail and Pattenson delivered an updated version of the frequency and signal control in which all the rotating parts and mechanical contacts were replaced by solid state logic circuitry. At the time of writing the control was a cesium standard, with a rubidium standard and one of the Sulzer crystals as an ultimate standby.

S.N. Kalra, R. Bailey, H. Daams, and Brian Rafuse of NRC completed construction of a long beam cesium standard in 1958. A 2 kc signal by telephone line from the transmitter was referred to it directly, so that frequency was maintained close to the desired value. In 1961 it was noted that the Dominion Observatory was responsible for time, NRC was responsible for frequency, and DOT maintained and serviced the transmitters. The rather specialized equipment which controlled the signal pulses, such as frequency standards, synthesizers, decade counters, and so on were the responsibility of the Dominion Observatory.

In 1959 it became possible to install another high-powered transmitter, and it was decided to place it on 14,670 kHz, leaving 3330 kHz on low power for local distribution. Two interesting events followed. It had been on the air for only a short while when the top of the wooden mast supporting one end of the doublet antenna caught fire and burned like a torch. It required a rigger to square off the top and readjust the guy wires, and feed to the antenna had to be tuned more carefully. When increased power was found to cause some interference to a station in Peru, a vertically supported dipole with a corner reflecting system (Sterba Curtain) was designed by W.A. Cumming of NRC to give a lobe from Ottawa in the direction 350° to serve northwestern Canada.

It became apparent, as new transmitters with increased power were placed in service, that the antenna system was in need of review. Because they were horizontal their radiation pattern was restricted to one path, while at right angles very little signal could be detected. For maximum usefulness, the signal should radiate uniformly in all directions, as with a vertical antenna.

But there was a more serious problem. The antenna is one of the elements which must be included when tuning the transmitter. And if it becomes detuned, as a horizontal one will due to an ice load which brings it closer to the ground, or due to wind and other meteorological causes, then heavy currents may develop which will damage other components. The alternatives were either to operate at reduced power or to install rigid vertical antennas with a properly designed coaxial feeder for each frequency. The first of these efficient units, together with a 20-kilowatt transmitter, was installed in 1967 by Technical Material Corporation (TMC) of New York. By the end of 1969 all three frequencies were radiating from vertical antennas with marked improvement in efficiency. Prior to this, the stray radiation at the site was strong enough that some of the sensitive equipment, such as the rubidium atomic standard and standby quartz frequency standard, had to be housed in one of the cottages in a room shielded with copper mesh screen. Now, with local radiation reduced to a minimum, the screen cubicle with its contents was installed in the main transmitter building and the cottage removed. Obsolence, which overtakes all such equipment, was recognized by NRC in 1971 with the acquisition of two more TMC transmitters. There were in all six transmitters, three for normal operation on each of the three CHU frequencies, and three standby units which could be switched on for necessary maintenance, all of them of TMC manufacture. There was a vertical antenna for each frequency and a dummy load that could be used when testing any of the transmitters. A primary frequency standard provided all the frequencies, from the high carrier frequencies down to one pulse per second, so that the phase relationship referred to earlier was preserved. Solid state logic circuitry furnished all the sequences involved, such as omitted or elongated pulses or split pulses, while a speaking clock gave the minute-by-minute bilingual time announcement.

A growing need for precisely timed recordings of voice communications had led to the development in Canada of a new system of time distribution. 'There appears to be a wide requirement ... to legally index command and control communication system voice recordings. These include police, airport control, navigation control and railway dispatching. In addition there are tape recordings made by various government agencies, such as the Canadian Parliamentary Proceedings.' The task of developing a method of generating and distributing time signals in a form that can be encoded with the voice recording has been the subject of recent research at NRC. The new system, developed by Edmunde Newhall Associates following specifications outlined by NRC, involves a master digital clock regulated by the cesium atomic standard at NRC and a number of crystal-controlled secondary digital clocks distributed across the country wherever they are required. Each secondary unit has a free-running accuracy of 1 in 10^7 (about one hundredth of a second a day) and battery standby to bridge power failures. It is also equipped to dial in to the master unit at NRC Ottawa by direct distance dialing telephone line and to receive a signal which will correct it to the nearest millisecond. The telephone contact can be programmed to operate daily or at some longer interval, and it can also be initiated by push button control.

The serial time code in which the signals are sent, described technically as FSK (frequency shift keying), is an adaption of a system widely used for commercial data transmission and found to be very reliable. The message from the master clock at NRC includes the second, minute, hour, and day number of the year, all requiring but a few seconds of transmission time. Various safeguards are built into the field clock to prevent it from receiving an error due to noise or a transmission fault.

Satisfactory transmission tests of the FSK serial time code have been made by synchronizing a field clock at NRC through long distance telephone loops to Vancouver, Winnipeg, Toronto, Washington, Montreal, Moncton, Halifax, and St John's. Travel time of the signal over the eight loops is given in the accompanying table. A rough comparison of half the delay time with the distance traversed indicates that the speed of transmission of the message is considerably slower than the speed of light. It indicates that after update St John's will be 11 milliseconds slow compared to Ottawa, and Vancouver will be 17 milliseconds slow. These measurements were made using leased government lines. They may be different

Telephone loop delay from Ottawa		
Location	Delay (milliseconds)	Distance (miles)
St John's	22	1100
Halifax	13	600
Moncton	14	540
Washington DC	15	470
Montreal	3	105
Toronto	6	225
Winnipeg	19	1050
Vancouver	34	2200

NOTE: Distances are not exact transmission distances but straight-line estimates taken from a map. The delay from Ottawa to the field site is one-half the loop delay, e.g. 17 milliseconds to Vancouver.

on the public telephone network. A delay of nearly 300 milliseconds will be introduced if the link is by satellite, but this discrepancy would be readily detected with reference to a time signal such as CHU or WWV. The coded message is generated in FSK format using frequencies of 2725 and 2925 Hz. It can be recorded directly onto a tape on one channel, together with voice on an adjacent channel, and displayed on a digital clock at a later time as the tape is examined. In mid-summer 1976 the same code was introduced to the CHU seconds pulses from 31 to 39 inclusive each minute. Users within radio range of CHU were thus enabled to use the radio link in place of the telephone. It remained to be seen how widely this new system of time distribution would be adopted.

With this glimpse of the future we conclude the story of timekeeping in Canada. It has been a review of the response of government authorities to the time requirements of navigation, meteorology, surveying, frequency control, and other branches of scientific research. We have in general restricted the story to the period from about 1840 to 1973. During this era the astronomical pendulum achieved its finest performance as a timekeeper in the concept of the synchronome free pendulum. It was replaced by the products of the physics laboratory, first the quartz oscillator, then the atomic resonator. The definition of the fundamental unit of time, the second, has had two modifications in the last two decades, and is now defined in terms of the cesium atom.

Today, Canada's time service is based on the most sophisticated equipment to be found in any laboratory. The long beam cesium standard was designed and built at the National Research Laboratory, and is one of three operating in national laboratories. As a group these laboratory instruments stand out in a large population of commercially designed cesium 'clocks.' Time is made available to the Canadian public by every available means of communication, radio, TV, telephone; by seconds pulses, bilingual voice announcements, and by special code. Perhaps the best known outlet is the CBC's one o'clock signal at 'the beginning of the long dash,' and almost as well known internationally is the short-wave broadcast CHU.

Equally important is the contribution Canada is making in the realization of a world clock, which is co-ordinated at the Bureau International de l'Heure in Paris, France. In addition to the cesium standard in Ottawa, the astronomical observations made with the photographic zenith tube at Ottawa and at Calgary, form an important link in the evaluation of mean solar time by the same office. And on the instruction of the BIH, the insertion of a leap second into the atomic time, at present once a year, keeps the atomic clock in step with mean solar time.

Our story has included men of vision and ability who have been trail blazers in this area of Canadian history. Equally significant has been the co-operation that has existed between departments of the public service, and between the public and private sector in Canada. In all this, Canadians have benefited.

The search for perfection is not an end in itself. The demands of other branches of science, including space research, are pushing hard on the heels of the timekeeper. Even so, among the units of science, no other unit is known with greater precision than the unit of time.

The unit of time

There are three natural units of time: the tropical year, marked by the return of the seasons, the lunar month, marked by the return of the new moon, and the day, marked by passage of the sun. All other units, such as the hour, minute, and second, are man-made to meet the needs of society.

It might be noted in passing that the three natural units are not multiples of each other. The tropical year contains 365.2422 days, and the return of the new moon occurs on the average in 29.53 days, which is equivalent to 12.368 lunar months in the year. The ancients tried unsuccessfully to discover simple relationships, but the struggle with the calendar does not form part of this discussion.

What time is, of course, nobody knows. Fortunately the lack of a satisfactory definition of time is, for most purposes, no great handicap: all that is necessary is to be able to measure it. One attendant problem, of course, is the choice of a suitable unit of time. The unit has been based progressively on the day, the tropical year, and a natural transition period within the cesium atom.

For the measurement of time there was supplied, when the earth was created, an almost perfect clock. The earth in space is spinning on its axis, and it travels in an orbit, almost circular, about the sun. Since the easiest way to count the days is to note each time that the rotation brings the sun overhead, a solar day has been used as the ordinary unit of time. The introduction of clocks forced the adoption of a mean solar day, because apparent solar time, as displayed by a sundial, showed variations of as much as a quarter of an hour fast and slow in the course of a year.

The second, defined as 1/86,400 of the mean solar day, became the fundamental unit of time. Earth rotation is measured by noting the passage of stars across the meridian, and the resulting 'sidereal time' is translated into 'mean solar time' by a simple formula. The primary pendulum timekeepers in the observatory were thus sidereal. When properly adjusted they were sealed in air-tight containers and left to run undisturbed in a thermally controlled vault. Any residual rate and resulting clock error was determined from the transit observations. The mean solar time pendulums assumed a secondary role and were adjusted as necessary each day.

In 1939, H. Spencer Jones, the astronomer royal, confirmed a growing suspicion by showing from astronomical evidence that the earth's rotation was not uniform.[1] He argued that either the members of the solar system had conspired to change their predicted positions by amounts proportional to their orbital speeds, or else the earth had changed its period of rotation. Hence the unit of time, one of the fundamental constants, was not really a constant. So long as timekeeping to an accuracy of better than a part in 10^8 (about a millisecond a day) was still academic, the problem was not too serious. But by 1950 the problem could no longer be ignored. Crystal clocks had been developed to the point where they could be used to detect departures from uniformity.

Universal time (UT), which is mean solar time referred to the Greenwich meridian, is not uniform because of the following irregularities:
– A shift in the earth's crust relative to the direction in space of the earth's axis of rotation, known as polar wander and measured as a variation of latitude.
– A seasonal change, detected with the use of modern timekeepers, which is assumed to be periodic and hence is predictable. It is recognized as a slowing down of earth rotation in the spring near the time of the vernal equinox, followed by a speeding up in the autumn, and is perhaps due to meteorological causes.
– A gradual secular slowing down, amounting to about a millisecond per century, which has little impact on the present discussion.
– A random unpredictable variation in earth rotation which may be due to interaction between the core and the mantle of the earth. In 1972 the period of earth rotation was about 3 milliseconds longer than its average period during the eighteenth and nineteenth centuries. There is no way of predicting when this slowing trend will reverse.[2]

Polar wander had been recognized since the turn of the century. About 1905, five small observatories were established as uniformly as possible

1 H.S. Jones, 'The rotation of the earth'
2 M.M. Thomson, *J. Roy. Astron. Soc. Can.*, 64 (1970)

about the world at 39°08′ north latitude. Using the same catalogue of stars and visual zenith telescopes of the same design, they recorded the cyclical variation of latitude of each station, thereby indicating the irregular motion of the pole of rotation with respect to its mean position. The principal period of polar wander is fourteen months. A result of the motion of the pole is an east-west shift of the meridian of each place, with a value of zero at the equator and becoming maximum for places at the north and south poles. The displacement is small, amounting at Ottawa to a maximum error of ±30 milliseconds of time (±3 feet displacement).

Modern observations of earth rotation on any one night can be made with an uncertainty of three or four milliseconds, and by averaging over a few nights can be reduced to a millisecond or less. The mean solar time derived from this smoothing is called UT0. The Bureau International de l'Heure in Paris provides to each of the national time services corrections for polar wander and seasonal variation which result in the following: UT0 plus correction for polar wander gives UT1; UT1 plus correction for seasonal variation gives UT2. UT2 is reasonably uniform from day to day, but the inherent irregularities due in particular to the random variations mentioned above make it quite inadequate for the definition of the unit of time, the second. However it must be noted that an accurate knowledge of UT is essential in all modern endeavours which require a knowledge of the attitude of the earth in space, such as surveying, celestial navigation, and the launching and control of space vehicles.

Responding to the general need for a revision in the unit of time, the Conference of Fundamental Constants, held in Paris in 1950 under the chairmanship of André Danjon, director of the Paris Observatory, proposed that the second be defined in terms of the year instead of the day, in particular the year 1900. The International Astronomical Union (IAU) at its General Assembly in Rome in 1952 gave its approval. Action on a new definition of the unit of time had to await a formal pronouncement by the International Committee of Weights and Measures in Paris. It was given in 1956 in the following resolution: 'The second is the fraction 1/31,556,925.9747 of the tropical year 1900, January 0, 12 hours Ephemeris Time.' This resolution was formally approved by the IAU in 1958. The reason for stating the year in this manner is that it defines the second as a constant. The tropical year is decreasing at the rate of 0.530 ephemeris second per century. It is thus easy to relate any particular year to the tropical year for 1900.

Ephemeris time (ET) is the independent variable in the equations which describe the motions of the sun, moon, and planets. It is essential to astronomy because it forms the continuous link between observations of the remote past and predictions into the future. Furthermore it can be observed directly by observing the position of any member of the solar system against the background of stars, and it is independent of earth rotation. The usual practice is to observe the moon as the time indicator and the background of stars as the celestial clock. Its position can be determined by the use of transit instruments, moon cameras, and by the observation of lunar occultations (see pp. 98–9).

The obvious practical problem is that ephemeris time is not readily available. The moon does move faster than the sun or any of the planets against the star background. Even so it requires $27^1/_2$ days to complete one circuit. The position of the moon in its orbit is predicted according to ephemeris time, but the observations are recorded in UT. The two times differ because ET gains on UT because the earth is rotating at reduced speed. Hence the apparent position of the moon is in error, both in right ascension and declination. This may be translated into the increment of time required for the moon to travel this small distance in each co-ordinate along its orbit. It is the difference ΔT between ephemeris time and universal time, and is expressed in the form ET = UT + ΔT. Observations must be averaged over a period of about a year to obtain a determination of ET with a precision approaching a part in 10^9. As the time base is expanded to a century, the relative precision approaches about 3 parts in 10^{10}.[3] While this is invaluable to the astronomer in relating events of the past and projecting forecasts into the future, it fails as a readily available time scale of short-term high precision for laboratory purposes.

The cesium resonator provided the solution to this problem. The first cesium atomic resonator, as we have seen, was placed in operation in 1955.[4] It was then calibrated with respect to the value of the second of UT2. After a three-year experiment conducted jointly between the National Physical Laboratory in England and the US Naval Observatory, the value of the ephemeris second was derived as 9,192,631,770 ± 20 cycles of a transition of the cesium atom, and this was also adopted by the International Astronomical Union in Moscow in 1958. But the proposal to adopt it as an alternative to the ephemeris second was rejected. Promising research with other elements, principally hydrogen and thallium, raised the question whether cesium would ultimately show the greatest precision and be readily reproduced at each repetition of the experiment.

3 Jean Kovalevsky, 'Astronomical time'
4 L. Essen and J.V.L. Parry, 'The cesium resonator as a standard of frequency and time'

Further research into the cesium resonator resulted in a greatly improved accuracy and reproducibility of laboratory standards, and also led to the introduction in quantity of commercially designed atomic clocks of high precision. Time could be intercompared between laboratories to the microsecond or better, and frequency to a part in 10^{11}. From the standpoint of the physicist all the requirements of high accuracy and of long term continuity were being demonstrated by the cesium atomic clock. The passage of years confirmed the original value, and in October 1967, the General Conference on Weights and Measures (CGPM) defined the second in the International System of Units as 9,192,631,770 periods of the radiation emitted by the transition between two hyperfine states of the cesium 133 atom in the ground state. At the same time the CGPM recognized the continuing use of the ephemeris second in all astronomical computations.

There was a continuing requirement for mean solar time in civil activities such as surveying, celestial navigation, and space exploration. Initially this was met by offsetting the output of the cesium standard to produce UT2. The offset was established on 1 January for the ensuing year, and any departure due to a change in earth rotation was accommodated by a step adjustment of a fraction of a second at the beginning of a month. The time and frequency produced in this manner were co-ordinated through the Bureau International de l'Heure and called Universal Time Co-ordinated (UTC).

As the table indicates, the offset in frequency commenced in 1961, and continued for eleven years. The system was inadequate because, while it indicated UT with but small error, it did not make the unit of time readily available to the scientist.

On 1 January 1972 the offset was removed from UTC, and the atomic second became the practical unit of time. The UTC clock now runs fast with respect to UT1, gaining about three milliseconds a day, or a little more than a second a year. Should the rate of earth rotation accelerate, it is quite possible for the UTC clock to lose with respect to UT1. In any case, provision has been made to adjust UTC by the addition or subtraction of a leap second at appropriate intervals so that at no time will UT1-UTC exceed seven-tenths of a second. The dates for applying the leap second are 1 July or 1 January at zero hours UT. Since UT1 is running slow, leap seconds must be inserted so that UTC will be retarded. This was done 1 July 1972 and 1 January 1973, and present indications are that leap seconds will be added annually for the next few years.

A paper value of atomic time, called International Atomic Time, (IAT)

Offsets and step adjustments of UTC until 31 December 1973

Date (at oh UT)	Offsets	Steps
1961 Jan. 1	-150×10^{-10}	
Aug. 1	"	$+0.050^s$
1962 Jan. 1	-130×10^{-10}	
1963 Nov. 1	"	-0.100^s
1964 Jan. 1	-150×10^{-10}	
Apr. 1	"	-0.100^s
Sep. 1	"	-0.100^s
1965 Jan. 1	"	-0.100^s
Mar. 1	"	-0.100^s
Jul. 1	"	-0.100^s
Sep. 1	"	-0.100^s
1966 Jan. 1	-300×10^{-10}	
1968 Feb. 1	"	$+0.100^s$
1972 Jan. 1	0	$-0.107\,7580^s$
Jul. 1	"	-1^s
1973 Jan. 1	"	-1^s

was extended back to 1 January, zero hours, 1958. At the end of December 1971, UTC was just a little less than ten seconds slow with respect to IAT. In order to make it ten seconds exactly, an increment of 0.107758 second was added to UTC. Each successive leap second has therefore resulted in the difference IAT-UTC being an integral number of seconds.

The difference, DUT1 = UT1-UTC, is indicated to the nearest tenth of a second by emphasizing certain seconds pulses on the radio time signals. On the signal emitted by the National Research Council of Canada time transmission, CHU, the emphasis is achieved by removing the central third of a pulse which is normally 0.3 second long, so that it sounds like a double pulse. If UT1 is fast appropriate pulses between seconds 1 and 7 are emphasized, and if UT1 is slow seconds 9 to 15 are used. For instance, if UT1 is ahead of UTC by two-tenths of a second, then pulses 1 and 2 will sound like double pulses, while if UT1 is slow by three-tenths of second, then pulses 9, 10, and 11 will be emphasized. The zero pulse of each minute is always lengthened and is never used for this program. Its leading edge marks the beginning of the minute.

The present system for the radio transmission of time signals makes the official unit of time instantly available. For some types of field work, the transmitted time is a close enough approximation to mean solar time

(UT1). The correction, DUT1, can be applied, yielding UT1 to the nearest tenth of a second. Some time transmissions have a further incoded correction which can be applied to yield UT1 to 0.02 second. Precision of this degree or better can, for most purposes, await the published corrections to the signals.

Daylight Saving Time

It was not until the beginning of the second world war that the adoption of daylight saving time was seriously considered, although propaganda had been carried on for some years previously. In 1916 an Act of Parliament legalized its use in Great Britain, and since that time it has been maintained there. The United States adopted daylight saving over the entire country in 1918, and in 1919 its use there was fairly general. Now, by the Uniform Time Act of 1966, the entire USA, with the exception of any State which exempts itself, observes daylight saving time each year from the last Sunday in April to the last Sunday in October.

In Canada, the Dominion Government, by the Daylight Saving Act of 1918, required the adoption of daylight saving during the summer of that year. After 1918 the Act lapsed.

During the second world war there were three orders-in-council passed under the War Measures Act dealing with the observance of daylight saving time:

i/ PC4494, 20 September 1940, directed that 'Daylight Saving Time, so called, (being one hour in advance of Standard Time) observed during the past summer months, shall continue to be observed until such time as the Governor-in-Council may otherwise order in the Province of Quebec and in the Province of Ontario by all persons, firms and corporations resident or carrying on business therein except transportation companies and telegraph companies.'

ii/ PC547, 26 January 1942, ordered that 'as of and from 2.00 A.M. Standard Time, Monday, February 9th 1942, until otherise ordered, the time for all purposes in Canada shall be one hour in advance of accepted Standard Time, and that Daylight Saving Time shall be observed by all persons, firms, corporations and public authorities, without exception, situate, resident or carrying on business in Canada.'

iii/ PC6102, 14 September 1945, revoked the above.

Several provinces have legislation controlling provincial or municipal adoption of daylight saving. In recent years, mainly through the impact of radio and television, and perhaps influenced by legislation within the USA, it has become an established custom in Canada, where daylight saving is observed, to advance the clocks for the six-month interval from the last Sunday in April to the last Sunday in October.

Bibliography

Bignell, Nancy. 'Official time signal: 100 years.' *McGill News*, Summer 1962

Bodily, L.N. and R.C. Hyatt. 'Flying clock experiments extended to East Europe, Africa, and Australia.' *Hewlett-Packard Journal*, December 1967

Bond, W.C. 'The spring-governor (drum chronograph).' *Annals of the Astronomical Observatory of Harvard College*, Vol. 1, Pt 11, p. 5. [Harvard College Archives]

Bourne, B.E. and D.W.R. McKinley. 'A pulsed crystal oscillator range calibrator,' NRC no. 1822, Oct. 1948

– 'A precision radio time signal,' NRC no. 2513, 1951

Brealey, G.A. 'Digital velocity generating computer for the Dominion Observatory mirror transit telescope.' *Pub. Dom. Obs.*, 25, no. 6 (1964)

– and R.W. Tanner. 'Photographic registration of transits and reduction of observations of the Ottawa mirror transit telescope.' *Pub. Dom. Obs.*, 25, no. 3 (1963)

Brock, T.L. 'HM Dockyard, Kingston.' *Historic Kingston, Transactions of the Kingston Historical Society*, no. 16 (1968)

Burland, Miriam S. 'Robert Meldrum Stewart.' *J. Roy. Astron. Soc. Can.*, 49, no. 2 (1955)

Burpee, L.J. *Sandford Fleming*. Oxford University Press, 1915

Carpmael, James and C.H. McLeod. 'The Longitude of the Toronto Observatory.' *Trans. Roy. Soc. Can.*, Sect. III (1888)

Comité International de Poids et Mesures. *Procès-verbaux des séances*, 2e sér 1957, xxv, 77 [definition of Ephemeris Time]

Cooke, W.E. and F.B. Cooke. 'Coincidence of wireless time signals by extinction.' *Monthly Notices Roy. Astron. Soc.*, 77 (1917)

Craig, J.D. 'Dr Edouard Gaston Deville.' *J. Roy. Astron. Soc. Can.*, 18, no. 10 (1924)

Davey, Ed. 'Reminiscences.' A letter to W.J. Wilson, Department of Communication, 18 Dec. 1967

Dennis, John S. jr. 'A short history of the surveys made under the Dominion Lands system 1869–89.' Canadian *Sessional Papers* no. 13, A 1892 (pt VI), Department of the Interior

Douglas, A. Vibert. 'Astronomy at Queen's University.' *J. Roy. Astron. Soc. Can.*, 52, no. 2 (1958)

– 'Early Canadian Astronomy.' *Queen's Quarterly*, 78, no. 4 (1971)

Dowd, Charles N. *Charles F. Dowd.* NY: Knickerbocker Press, 1930

Essen, L. and D.W. Dye. [Description of ring crystal oscillator]. IEE Proceedings, Pt 2 (1951)

Essen, L. and J.V.L. Parry. 'The cesium resonator as a standard of frequency and time.' *Phil. Trans. Roy. Soc. Lond.*, Series A, 250 (1957)

Essen, L., J.V.L. Parry, W. Markowitz, and R.G. Hall. 'The value of the ephemeris second in terms of the cesium atom.' *Nature* [London] 181, 1054 (1958)

Explanatory Supplement to the Ephemeris. London: HMSO 1961

Fleming, (Sir) Sandford. *Uniform non-local time.* Published privately (1876). [Public Archives of Canada]

– 'Time reckoning for the twentieth centure.' *Trans. Roy. Soc. Can.*, Sec. III, (1888)

– 'The unit measure of time' [Presidential address]. *Trans. Roy. Soc. Can.*, Sec. III (1890)

Grant, G.M. *From ocean to ocean. Sandford Fleming's expedition through Canada in 1872.* Toronto: James Campbell and Son, 1873.

Harvard College. *Report of the director for the year 1852.* [Harvard University Archives]

Henderson, J.P. 'Daily journal and diary, 1919–56.' [A record of daily activities at the Dominion Observatory, as well as his own private activities]

– 'Wireless in the Mackenzie Basin.' *J. Roy. Astron. Soc. Can.*, 16, no. 2 (1922)

– 'Time comparison by coincidence.' *J. Roy. Astron. Soc. Can.*, 19, no. 2 (1925)

– 'Wireless time signals and their applications.' *J. Roy. Astron. Soc. Can.*, 22, no. 3 (1928)

Hollinsworth, V.E. 'Daily Journal.' [His activities at the Observatory; a few yearly journals preserved in possession of J.P. Henderson.]

– 'A program machine for automatic operation of the Ottawa PZT.' *Pub. Dom. Obs.* 15, no. 4 (1955)

– 'On the use of crystal controlled synchronous motors for the accurate measurement of time.' *Can. J. Res.*, 27 F (1949)

Hope-Jones, F. *Electrical timekeeping.* London: NAG Press, 1940

Jack Papers. Correspondence of W. Brydone Jack, Fredricton, with Sir George B. Airy, the astronomer royal, and W.C. Bond, director of Harvard College Observatory. Harriet Irving Library, University of New Brunswick

Jones, H. Spencer. 'The rotation of the earth and the secular accelerations of the sun moon and planets.' *Monthly Notices Roy. Astron. Soc.*, 99, (1939) 541

Journal of the House of Assembly of Nova Scotia, 1832

Kalra, S.N., R. Bailey, and H. Daams. 'Canadian cesium beam standard of frequency.' *Nature*, 183, (28 Feb. 1959) 575

– 'Cesium beam standard of frequency.' *Can. J. Phys.*, 36, (1958) 1442

– Pattenson, C.F. and M.M. Thomson. 'Canadian standard of frequency.' *Can. J. Phys.*, 37, (1959) 10

Kennedy, J.E. 'Plaque unveiled on the first astronomical observatory in Canada'; 'The early days of the first astronomical observatory in Canada.' *J. Roy. Astron. Soc. Can.*, 49 (1955)

– 'Our heritage in Canadian astronomy.' *J. Roy. Astron. Soc. Can.*, 66 (1972)

Klotz, J. Otto. 'Diary.' MG 30, C1, Public Archives of Canada, Ottawa

– 'Observatories in Canada (historical).' *J. Roy. Astron. Soc. Can.*, 12, no. 5 (1918); 13, no. 1 (1919); 13, no. 7 (1919)

– 'Astronomy, wireless and other.' *J. Roy. Astron. Soc. Can.*, 15 (1921) 306

Kovalevsky, Jean. 'Astronomical time.' *Metrologia* 1 (1965)

Logan, W.E. *Report of Progress*. Geological Survey of Canada, 1857

Markowitz, W. 'Dual Rate Moon Camera.' *Astron. J.*, 52 (1954)

Mayall, R. Newton. 'The inventor of standard time.' *Pop. Astron.*, 50, no. 4 (april 1942)

MacKinnon, K.A. 'Carrier frequency control.' *Electr. Dig.* (Jan. 1936)

McAlpine's Maritime Directory for 1870. Public Archives of Canada

McClenahan, W.S. 'Catalogue of 2436 stars from observations with the reversible meridian circle made at the Dominion Observatory Ottawa, during the years 1911–23.' *Pub. Dom. Obs.*, 15, no. 1 (1952)

– 'Results of observations made with the reversible meridian circle, 1923–35; Observations of the Sun, Mercury and Venus; catalogue of 1589 stars; Corrections to GC and FK3.' *Pub. Dom. Obs.*, 15, no. 2 (1952)

– E.G. Woolsey, and R.W. Tanner. 'Results made with the reversible meridian circle, 1935–50; catalogue of 1525 stars; Corrections to GC and FK3.' *Pub. Dom. Obs.*, 15, no. 3 (1954)

McDiarmid, F.A. 'Determination of the 141st meridian.' *J. Roy. Astron. Soc. Can.*, 2 (1908) 93

McGill University Scrap Book. McGill University Archives

McGill University, 'Governor's minute book.' McGill University Archives

Miller, A.H. 'Dominion Observatory work in the Mackenzie District.' *J. Roy. Astron. Soc. Can.*, 17, no. 5 (1923)

Mungall, A.G. 'Atomic time scales.' *Metrologia*, 7 (1971)

Mungall, A.G., R. Bailey, and H. Daams. 'Atomic transition sets Canada's time.' *Can. Electr. Eng.*, (August 1965)
- D. Morris, R. Bailey, and H. Daams. 'Atomic hydrogen maser development.' *Metrologia*, 4 (1968)
- R. Bailey, and H. Daams. 'The Canadian cesium beam frequency standard.' *Metrologia*, 2, (1966)
- R. Bailey, H. Daams, D. Morris, and C.C. Costain. 'The new 2.1 metre primary cesium beam frequency standard, CsV.' *Metrologia*, 9 (1973)

Nugent, D.B. 'Charles Campbell Smith.' *J. Roy. Astron. Soc. Can.*, 34 (1940)

Orders-in-council: PC 4494, 20 Sept. 1940, daylight saving time imposed in Ontario and Quebec
- PC 6784, 28 Aug. 1941, Dominion Observatory time declared official for official purposes
- PC 547, 26 Jan. 1942, daylight saving imposed Canada-wide
- PC 6102, 14 Sept. 1945, federal daylight saving law rescinded
- PC 1970-562, 26 March 1970, transferred authority from the Dominion Observatory to the National Research Council of Canada

Patterson, J. 'A century of Canadian meteorology.' *Q. J. Roy. Meteor. Soc.*, 66, no. 16 (1940)

Plaskett, J.S. 'William Frederick King.' *J. Roy. Astron. Soc. Can.*, 10, no. 6 (1916)

Queen's University Minute Book, 1905–6–8–12. Queen's University Archives

Raddall, T.H. *Halifax, Warden of the north.* Toronto: McClelland and Stewart, 1948

Raymond, W.O. (Archdeacon). 'Genesis of the University of New Brunswick.' (1918) Harriet Irving Library, University of New Brunswick

Rogers, W.A. and C.H. McLeod. 'The longitude of McGill Observatory.' *Trans. Roy. Soc. Can.*, Sec. III (1885)

Scadding, Henry. *Toronto of old.* Toronto: Adam, 1873

Sessional Papers, Department of Marine and Fisheries, 1868 to 1936 [contain the annual reports of the directors of the observatories at Toronto, Montreal, Quebec, Saint John, Kingston, Vancouver, and Victoria]

Sessional Papers, Department of the Interior, 1892 to 1937, [contain annual reports of the astronomical work of the Department]

Smith, C.C. 'Activities of the Astronomical Branch (history).' Department of the Interior, 1930, Vol. 2, Sec. A, Division of Positional Astronomy
- 'Converting a transit from a straight to a broken type.' *J. Roy. Astron. Soc. Can.*, 28 (1934)

Stewart, R.M. 'Report of the superintendent of the time service.' *Report of the chief astronomer 1908*, Appendix 3
- 'The time service at the Dominion Observatory.' *J. Roy. Astron. Soc. Can.*, 1 (1907)

- 'A new form of clock synchronism.' *J. Roy. Astron. Soc. Can.*, 7 (1913)
- 'Wireless time signals observed at Ottawa.' *J. Roy. Astron. Soc. Can.*, 19 (1925)
- 'Comparison of wireless time signals from Annapolis and Lafayette.' *J. Roy. Astron. Soc. Can.*, 16 (1922)
- 'The northern belt of international longitudes.' *Dom. Obs.* reprint no. 26 (1934)
- 'Dr J. Otto Klotz.' *J. Roy. Astron. Soc. Can.*, 18 (1924)
- 'Activities of the Astronomical Branch (history).' Department of the Interior, 1930, Vol. 1. Dominion Observatory Pamphlet M2

Swail, J.C. and C.F. Pattenson. 'A carrier frequency and time signal control system.' *J. Sci. Instr.*, 40, (June 1963)

Tanner, R.W. 'UT, UTC, ET, and AT.' *J. Roy. Astron. Soc. Can.*, 61 (1967)
- 'Method and formula used in the photographic zenith tube plate measurement.' *Dom. Obs. Pub.* 15, no. 4 (1955)

Thomas, D.V. 'Photographic zenith tube instrument and method of reduction.' *Obs. Bull.* no. 81, London: HMSO 1964

Thomson, Don W. *Men and Meridians.* Vol. 1, prior to 1867 (1966); Vol. 2, 1867 to 1917 (1967); Vol. 3, 1917 to 1947 (1969). Ottawa: Queen's Printer

Thomson, M.M. 'Canada's time service.' *J. Roy. Astron. Soc. Can.*, 42 (1948)
- 'Markowitz dual-rate moon camera.' *J. Roy. Astron. Soc. Can.*, 52 (1958)
- 'Standard time and time zones in Canada.' *J. Roy. Astron. Soc. Can.*, 64 (1970)
- 'The Ottawa Photographic Zenith Tube.' *Pub. Dom. Obs.*, 15, no. 4 (1955)
- 'Private diary,' 1930–72
- 'A projection type measuring engine for photographic zenith tube plates.' *Pub. Dom. Obs.*, 15, no. 5 (1956)

Watts, C.B. 'The marginal zone of the moon.' *Astron. Pap. Amer. Ephem.*, 17, no. 9 (1963)

White, G.A. *Halifax and its Business.* Halifax: G.A. White Print Co. 1876

Williamson, Rev. James A. 'Report to the Board of Visitors by the director of the Kingston observatory for the year ending 31st December, 1864–5–6–7.' Queen's University Archives

Winkler, G.R.M., R.G. Hall, and D.B. Percival. 'The US Naval Observatory clock time reference and the performance of a sample of atomic clocks.' *Metrologia*, 4 (1970)

Woolsey, E.G. and R.W. Tanner. 'Results of observations made with the reversible meridian circle 1950–53; catalogue of 812 stars; corrections to 347 FK3 stars.' *Pub. Dom. Obs.* 15, no. 7 (1956)

Woolsey, E.G. 'Results of observations made with the meridian circle 1956–61; catalogue of 3753 AGK3 reference stars; corrections to 930 FK3 stars.' *Pub. Dom. Obs.*, 25, no. 8 (1966)

Index

Date Due